Multi-Objective Optimization in Computer Networks Using Metaheuristics

OTHER TELECOMMUNICATIONS BOOKS FROM AUERBACH

Multi-Objective Optimization in Computer Networks Using Metaheuristics

Yezid Donoso
Ramon Fabregat

CRC Press
Taylor & Francis Group
Boca Raton London New York

CRC Press is an imprint of the
Taylor & Francis Group, an **informa** business
AN AUERBACH BOOK

First published 2007 by Auerbach Publications

Published 2019 by CRC Press
Taylor & Francis Group
6000 Broken Sound Parkway NW, Suite 300
Boca Raton, FL 33487-2742

© 2007 by Taylor & Francis Group, LLC
CRC Press is an imprint of the Taylor & Francis Group, an informa business

No claim to original U.S. Government works

ISBN-13: 978-0-8493-8084-6 (hbk)

Library of Congress Cataloging-in-Publication Data
Donoso, Yezid.
Multi-objective optimization in computer networks using metaheuristics / Yezid Donoso, Ramon Fabregat.
p. cm.
Includes bibliographical references and index.
ISBN 0-8493-8084-7 (alk. paper)
1. Computer networks. 2. Mathematical optimization. I. Fabregat, Ramon, 1963- II. Title.
TK5105.5.D665 2007
004.6--dc22   2007060385

Visit the Taylor & Francis Web site at
http://www.taylorandfrancis.com

and the Auerbach Web site at
http://www.auerbach-publications.com

Dedication

To my wife, Adriana

For her love and tenderness and for our future together

To my children, Andres Felipe, Daniella, Marianna,

and the following with Adry

… a gift of God to my life

Yezid

To my wife, Telvys, and my children

… "continuarem caminant cap a Ìtaca"

Ramon

Contents

Preface

Many new multicast applications emerging from the Internet, such as Voice-over-IP (VoIP), videoconference, TV over the Internet, radio over the Internet, video streaming multipoint, etc., have the following resource requirements: bandwidth consumption, end-to-end delay, delay jitter, packet loss ratio, and so forth. It is therefore necessary to formulate a proposal to specify and provide the resources necessary for these kinds of applications so they will function properly.

To show how these new applications can comply with these requirements, the book presents a multi-objective optimization scheme in which we will analyze and solve the problems related to resources optimization in computer networks. Once the readers have studied this book, they will be able to extend these models by adding new objective functions, new functions that act as restrictions, new network models, and new types of services or applications.

This book is for an academic and scientific setting. In the professional environment, it is focused on optimization of resources that a carrier needs to know to profit from computer resources and its network infrastructure. It is very useful as a textbook mainly for master's- or Ph.D.-level courses, whose subjects are related to computer networks traffic engineering, but it can also be used for an advanced or specialized course for the senior year of an undergraduate program. On the other hand, it can be of great use for a multi-objective optimization course that deals with graph theory by having represented the computer networks through graphs.

The book structure is as follows:

> Chapter 1: Analyzes the basic optimization concepts, as well as several techniques and algorithms for the search of minimals.

Chapter 2: Analyzes the basic multi-objective optimization concepts and the ways to solve them through traditional techniques and several metaheuristics.

Chapter 3: Shows how to analytically model the computer network problems dealt with in this book.

Chapter 4: The book's main chapter — it shows the multi-objective models in computer networks and the applied way in which we can solve them.

Chapter 5: An extension of Chapter 4, applied to optical networks.

Chapter 6: An extension of Chapter 4, applied to wireless networks.

Lastly, Annex A provides the source code to solve the mathematical model problems presented in this book through solvers. Annex B includes some source codes programmed in C language, which solve some of the multi-objective optimization problems presented. These source files are available online at http://www.crcpress.com/e_products/downloads/default.asp

The Authors

Yezid Donoso, Ph.D., is a professor at the Universidad del Norte in Barranquilla, Colombia, South America. He teaches courses in computer networks and multi-objective optimization. He is also a director of the computer network postgraduate program and the master program in system and computer engineering. In addition, he is a consultant in computer network and optimization for Colombian industries. He earned his bachelor's degree in system and computer engineering from the Universidad del Norte, Barranquilla, Colombia, in 1996; M.Sc. degree in system and computer engineering from the Universidad de los Andes, Bogotá, Colombia, in 1998; D.E.A. in information technology from Girona University, Spain, in 2002; and Ph.D. (cum laude) in information technology from Girona University in 2005.

Dr. Donoso is a senior member of IEEE as well as a distinguished visiting professor (DVP) of the IEEE Computer Society. His biography has been published in *Who's Who in the World* (2006) and *Who's Who in Science and Engineering* (2006) by Marquis, U.S.A. and in *2000 Outstanding Intellectuals of the 21st Century* (2006) by International Biographical Centre, Cambridge, England. His awards include the title of distinguished professor from the Universidad del Norte (October 2004) and the National Award of Operations research from the Colombian Society of Operations Research (2004).

Ramon Fabregat, Ph.D., earned his degree in computer engineering from the Universitat Autónoma de Barcelona (UAB), Spain, and his Ph.D. in information technology (1999) from Girona University, Spain. Currently, he is a professor in the electrical engineering, computer science, and automatic control departments and a researcher at the Institute of Informatics and Applications at Girona University. His teaching duties include graduate- and postgraduate-level courses on operating systems, computer communication networks, and the performance evaluation of telecommunication systems. His research interests are in the fields of management and performance evaluation of communication networks, network management based on intelligent agents, MPLS and GMPLS, and adaptive hypermedia systems.

He coordinated the participation of broadband communications and distributed systems research group (BCDS) in the ADAPTPlan project (a Spanish national research project). He is a member of the Spanish Network of Excellence in MPLS/GMPLS networks, which involves several Spanish institutions. He has participated in the technical program committees of several conferences and has coauthored several papers published in international journals and presented at leading international conferences.

Chapter 1

Optimization Concepts

In the field of engineering, solving a problem is not enough; the solution found must be the best solution possible. In other words, one must find the optimal solution to the problem. We say that this is the best possible solution because in the real world this problem may have certain constraints by which the solutions found may be feasible (they can be implemented in practice) and unfeasible (they cannot be implemented).

In engineering, one speaks of optimization when one wants to solve complex problems. Such complexity may be associated with the kind of problem one wants to solve (i.e., if the problem is nonlinear) or the kind of solution one wishes to get (i.e., whether the solution is exact or an approximation).

There are five basic ways to solve such problems: analytically, numerically, algorithmically through heuristics, algorithmically through metaheuristics, or through simulation. Analytical solutions are practically possible for simple problems, but complex or large-sized problems are very difficult and require too much computational time. When the analytical model is very complex, problems can be solved by approximation using numerical methods. To obtain such optimal approximate values, functions analyzed must usually meet a series of conditions. If such conditions are not met, the numerical method may converge toward the optimal value. In any event, these techniques are very useful when the problems are mono-objective, whether linear or not.

However, when a problem is multi-objective, numerical methods are susceptible of being nonconvergent, depending on the model used. For example, if one attempts to solve a multi-objective optimization scheme by means of numerical methods using the mono-objective scheme of the weighted sum of the functions (which will be explained later), in addition to the conditions that the functions must meet for the specific numerical method, such a multi-objective scheme would present inconveniences if the search space is not convex, because one might not find many solutions. One can also find the solution applying computational algorithms called heuristics. In this case, heuristics presents a computational scheme that can reach the optimal value in an computational time. But the search of solutions with this type of algorithm may exhibit serious problems if, for example, the spaces are nonconvex or if the nature of the problem is of combined solution analysis. Furthermore, heuristics may present serious computational time problems when the problem is NP-Hard. A recognition problem P_1 is said to be NP-Hard if all other problems in the class NP polynomially to P_1, and we say that a recognition problem P_1 is in the class NP if for every yes instance of P_1, there is a short (i.e., polynomial length) verification that the instance is a yes instance [AHU93].

To overcome these inconveniences, metaheuristics have been created, which obtain an approximate solution to practically any kind of problem that is NP-Hard and complex or combined analysis solutions. Among existing metaheuristics we can mention genetic algorithms, Tabu search, ant colony, simulated annealing, memetic algorithms, etc. Many of these metaheuristics have been redesigned to provide solutions to multi-objective problems, which are the main interest of this book.

This chapter provides an introduction to fundamentals of local and global minimal and some existing techniques to search for such minimal.

1.1 Local Minimum

When optimizing a function $f(x)$, one wants to find the minimum value in an $[a, b]$ interval; that is, $a \leq x \leq b$. This minimum value is called the **local minimum** (Figure 1.1).

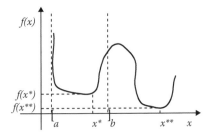

Figure 1.1 Local minimum.

If given function $f(x)$, we want to find the minimum value, but only in the $(a \leq x \leq b)$ interval. The resulting $f(x^*)$ value is called the local minimum of function $f(x)$ in interval $[a, b]$. As shown in Figure 1.1, this $f(x^*)$ value is the minimum value in the $[a, b]$ interval, but is not the minimum value of function $f(x)$ in the $(-\infty, \infty)$ interval.

Traditionally, search techniques for local minima are simpler than search techniques for global minima due, among many reasons, to the complexity generated in the search space when the interval is $(-\infty, \infty)$.

1.2 Global Minimum

When the function minimized is not constrained to a specific interval of the function, then one says that the value found is a **global minimum**. In this case, the search space interval is associated with $(-\infty, \infty)$.

Even though in Figure 1.1 the value of function $f(x^*)$ is a minimum, we can see that $f(x^{**}) < f(x^*)$. If there is no other value $f(x')$, so that $f(x') < f(x^{**})$ in the $(-\infty, \infty)$ interval, then one says that $f(x^{**})$ is a global minimum of function $f(x)$.

To find the **global maximum** value of a function, the analysis would be exactly the same, but in this case there should not exist another $f(x')$ value so that $f(x') > f(x^{**})$ in the $(-\infty, \infty)$ interval.

1.3 Convex and Nonconvex Sets

Definition 1

A set S of $\Re^n$ is **convex** if for any pairs of points P_1, $P_2 \, \varepsilon \, S$ and for every λ $\varepsilon \, [0, 1]$ one proves that $P = \lambda P_1 + (1 - \lambda)P_2 \, \varepsilon \, S$. Point P is a linear combination of points P_1 and P_2.

A set $S \subseteq \Re^n$ is convex if the linear combination of any two points in S also belongs to S.

On the other hand, a set is **nonconvex** if there is at least one point P in set S that cannot be represented by a linear combination.

Taking into account these definitions, Figure 1.2 represents convex solution sets and Figure 1.3 represents nonconvex solution sets.

Below are examples of convex sets.

Exercise

a. Is set $S = \left\{ (x_1, x_2) \in \Re^2 \ / \ x_2 \geq x_1 \right\}$ convex?

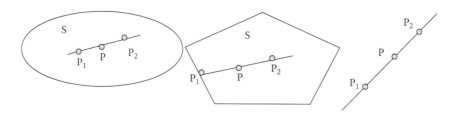

Figure 1.2 Convex sets.

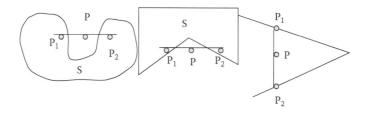

Figure 1.3 Nonconvex sets.

Proof:

Let $x = (x_1, x_2)$, $y = (y_1, y_2)$, and $x, y \, \varepsilon \, S$. We must prove that for every $\lambda \, \varepsilon [0, 1]$, $z = (z_1, z_2) = \lambda x + (1 - \lambda)y \, \varepsilon \, S$. In this case, we must prove that $z_2 \geq z_1$.

$$\lambda x + (1 - \lambda)y = (\lambda x_1 + (1 - \lambda)y_1, \lambda x_2 + (1 - \lambda)y_2)$$

Because $x, y \, \varepsilon \, S$, $x_2 \geq x_1$, $y_2 \geq y_1$, and $\lambda \geq 0$ y $(1 - \lambda) \geq 0$.

Then $\lambda x_2 \geq \lambda x_1$ y $(1 - \lambda)y_2 \geq (1 - \lambda)y_1$.

Adding both inequalities, we have

$$\lambda x_2 + (1 - \lambda)y_2 \geq \lambda x_1 + (1 - \lambda)y_1$$

Because $z \, \varepsilon \, S$, $z_2 = \lambda x_2 + (1 - \lambda)y_2$ y $z_1 = \lambda x_1 + (1 - \lambda)y_1$.

Replacing in the foregoing inequality we have that $z_2 \geq z_1$; therefore we have proven that S is a convex set.

b. Let $S = \left\{ x \in \Re \, / \, |x| \leq 1 \right\}$. Is it convex?

Proof:

Let x, $y \in S$, that is, $|x| \le 1$ and $|y| \le 1$. We must prove that for every $\lambda \in [0, 1]$, $z = \lambda x + (1 - \lambda)y \in S$. In this case, we must prove that $|z| \le 1$.

Because $z = \lambda x + (1 - \lambda)y$, applying $| |$ to the equality and taking into account the properties of this function — that $\lambda \in [0, 1]$, $|x| \le 1$, and $|y| \le 1$ — we have that $\rightarrow |z| = |\lambda x + (1 - \lambda)y| = |\lambda x| + |(1 - \lambda)y| = \lambda|x| + (1 - \lambda)|y| \le 1$, and therefore S is convex.

1.4 Convex and Concave Functions

Definition 2

Let S be a convex, unempty subset of $\Re^n$ and f a defined function of S in $\Re$. Function f is **convex** in S if and only if for any pair of points P_1, $P_2 \in S$ and for every $\lambda \in [0, 1]$ one proves that $f(\lambda P_1 + (1 - \lambda)P_2) \le \lambda f(P_1) + (1 - \lambda)f(P_2)$.

Definition 3

Let S be a convex, unempty subset of $\Re^n$ and f a defined function of S in $\Re$. Function f is **concave** in S if and only if for any pair of points P_1, $P_2 \in S$ and for every $\lambda \in [0, 1]$ one proves that $f(\lambda P_1 + (1 - \lambda)P_2) \le \lambda f(P_1) + (1 - \lambda)f(P_2)$.

Definition 4

Let S be a convex, unempty subset of $\Re^n$ and f a defined function of S in $\Re$. Function f is **strictly convex** in S if and only if for any pair of points P_1, $P_2 \in S$ and for every $\lambda \in [0, 1]$ one proves that $f(\lambda P_1 + (1 - \lambda)P_2) \le \lambda f(P_1) + (1 - \lambda)f(P_2)$.

Definition 5

Let S be a convex, unempty subset of $\Re^n$ and f a defined function of S in $\Re$. Function f is **strictly concave** in S if and only if for any pair of points P_1, $P_2 \in S$ and for every $\lambda \in [0, 1]$ one proves that $f(\lambda P_1 + (1 - \lambda)P_2) \le \lambda f(P_1) + (1 - \lambda)f(P_2)$.

Exercice

Probe whether the following functions are concave or convex.

a. Let $y = f(x) = ax + b$, in $\Re$, with $a, b \, \varepsilon \, \Re$.

   Proof:

   Let

   $$f(\lambda x + (1 - \lambda)y) = a[x + (1 - \lambda)y] + b$$

   $$= a\lambda x + a(1 - \lambda)y + b$$

   $$= \lambda ax + (1 - \lambda)ay + b + b - \lambda b$$

   $$= \lambda ax + b + (1 - \lambda)ay + b - \lambda b$$

   $$= \lambda[ax + b] + (1 - \lambda)ay + (1 - \lambda)b$$

   $$= \lambda[ax + b] + (1 - \lambda)[ay + b]$$

   $$= \lambda f(x) + (1 - \lambda)f(y)$$

   Consequently, function $y = f(x) = ax + b$ is convex and concave.

b. Let $y = f(x) = \displaystyle\sum_{i=1}^{n} c_i.x_i$, of $\Re^n$ in $\Re$, with c_i, x_i positive.

   Proof: Due to the function we are stating, in this case the inequality $<$ does not apply, because these types of functions are not convex.

   Let

   $$f(\lambda x + (1 - \lambda)y) \geq \sum_{i=1}^{n} c_i.\left(\lambda x_i + (1 - \lambda)y_i\right)$$

   $$\geq \sum_{i=1}^{n} \lambda c_i x_i + \sum_{i=1}^{n} (1 - \lambda)c_i y_i$$

$$\geq \lambda \sum_{i=1}^{n} c_i x_i + (1 - \lambda) \sum_{i=1}^{n} c_i y_i$$

$$\geq \lambda f(x) + (1 - \lambda) f(y)$$

Consequently, function $y = f(x) = \sum_{i=1}^{n} c_i x_i$ is concave.

The following definitions can be used when function f is differentiable.

Definition 6

Let S be a convex, unempty subset of $\Re^n$ and f a defined differential function of S in $\Re$. Then, function f is **convex** in S if and only if for any pair of points P_1, $P_2 \in S$ one proves that $\left[\nabla f(y) - \nabla f(x)\right]\left(y - x\right) \geq 0$.

Definition 7

Let S be a convex, unempty subset of $\Re^n$ and f a defined differential function of S in $\Re$. Then, function f is **strictly convex** in S if and only if for any pair of points P_1, $P_2 \in S$ one proves that $\left[\nabla f(y) - \nabla f(x)\right]\left(y - x\right) > 0$.

Definition 8

Let S be a convex, unempty subset of $\Re^n$ and f a defined differential function of S in $\Re$. Function f is **concave** in S if and only if for any pair of points P_1, $P_2 \in S$ one proves that $\left[\nabla f(y) - \nabla f(x)\right]\left(y - x\right) \leq 0$.

Definition 9

Let S be a convex, unempty subset of $\Re^n$ and f a defined differential function of S in $\Re$. Function f is **strictly concave** in S if and only if for any pair of points P_1, $P_2 \in S$ one proves that $\left[\nabla f(y) - \nabla f(x)\right]\left(y - x\right) < 0$.

Exercise

Show whether the following functions are concave or convex.

c. Let $y = f(x) = \sum_{i=1}^{n} x_i^2$, of $\Re^n$ in $\Re$.

Proof:

In this case we obtain the gradient vector of $f(x)$, which consists of the first derivates for $f(x)$.

$$\nabla f(x) = \begin{bmatrix} \dfrac{\partial f}{\partial x_1} \\ \dfrac{\partial f}{\partial x_2} \\ \dots \\ \dfrac{\partial f}{\partial x_i} \\ \dots \\ \dfrac{\partial f}{\partial x_n} \end{bmatrix} = \begin{bmatrix} 2x_1 \\ 2x_2 \\ \dots \\ 2x_i \\ \dots \\ 2x_n \end{bmatrix} = \begin{bmatrix} 0 \\ 0 \\ \dots \\ 0 \\ \dots \\ 0 \end{bmatrix}$$

Afterwards, we will calculate the Hessian matrix, which consists of the second derivates for function $f(x)$.

$$Hf(x) = \begin{bmatrix} \dfrac{\partial^2 f}{\partial x_1^2} & \dfrac{\partial^2 f}{\partial x_1 x_2} & \dots & \dfrac{\partial^2 f}{\partial x_1 x_n} \\ \dfrac{\partial^2 f}{\partial x_2 x_1} & \dfrac{\partial^2 f}{\partial x_2^2} & \dots & \dfrac{\partial^2 f}{\partial x_2 x_n} \\ \dots & & & \\ \dfrac{\partial^2 f}{\partial x_i x_1} & \dfrac{\partial^2 f}{\partial x_i x_2} & \dots & \dfrac{\partial^2 f}{\partial x_i x_n} \\ \dots & & & \\ \dfrac{\partial^2 f}{\partial x_n x_1} & \dfrac{\partial^2 f}{\partial x_n x_2} & \dots & \dfrac{\partial^2 f}{\partial x_n^2} \end{bmatrix} = \begin{bmatrix} 2 & 0 & \dots & 0 \\ 0 & 2 & \dots & 0 \\ \dots & & & \\ 0 & 0 & \dots & 0 \\ \dots & & & \\ 0 & 0 & \dots & 2 \end{bmatrix}$$

Because the Hessian matrix is positive, function $y = f(x) = \sum_{i=1}^{n} x_i^2$ is convex.

d. Let $f(x) = e^x$.

Proof:

Substituting the value of the function in definition $\left[\nabla f(y) - \nabla f(x)\right]\left(y - x\right)$ we get:

$$[\nabla f(y) - \nabla f(x)] \, (y - x) = [e^y - e^x] \, (y - x)$$

Because $x \neq y$, if $x > y$, we have $e^x > e^y$, and if $y > x$, we have $e^y > e^x$.

Then, the sign of $[e^{ye} - e^x]$ is the same sign of $(y - x)$, the product $[e^{ye} - e^x] \, (y - x)$ will always be positive, and therefore, function $f(x) = e^x$ is strictly convex.

e. Let $f(x) = \ln x$, with $x > 0$.

Proof:

Substituting the value of the function in definition $\left[\nabla f(y) - \nabla f(x)\right]\left(y - x\right)$ we get:

$$[\nabla f(y) - \nabla f(x)] \, (y - x) = [1/y - 1/x] \, (y - x)$$

$$= \frac{(x - y)(y - x)}{xy}$$

Because $x \neq y$, if $x > y$ or $y > x$, then the statement $\dfrac{(x - y)(y - x)}{xy}$ will always be negative, and therefore, function $f(x) = \ln x$ is strictly concave.

Proposition 1

Let $(g \circ f)(x)$ with $f: H \subset \Re^n \to M$ and g: $M \subset \Re \to \Re$; H is a convex subset of $\Re^n$ and M an interval of $\Re$. Then:

If f is convex and g is increasing and convex, then function $(g \circ f)(x)$ is convex.

If f is convex and g is increasing and concave, then function $(g \circ f)(x)$ is concave.

If f is concave and g is increasing and convex, then function $(g \circ f)(x)$ is convex.

If f is concave and g is increasing and concave, then function $(g \circ f)(x)$ is concave.

Exercise

Probe whether the following functions are concave or convex.

f.   Let $y = f(x) = \ln\left(\sum_{i=1}^{n} x_i\right)$, of $\Re^n$, with x positive.

Proof:

Because we have previously proven that function $\sum_{i=1}^{n} x_i$ is concave, because function $\ln x$ is increasing and concave, applying proposition 1, we have that $y = f(x) = \ln\left(\sum_{i=1}^{n} x_i\right)$ is a concave function.

g.   Let $y = f(x) = \left(\sum_{i=1}^{n} x_i^2\right)^2$, with $x \ \varepsilon \ \Re^n$.

Proof:

Because we have proven that function $\sum_{i=1}^{n} x_i^2$ is convex and function x^2 is an increasing convex function, by proposition 1, we have that $y = f(x) = \left(\sum_{i=1}^{n} x_i^2\right)^2$ is a convex function.

1.5 Minimum Search Techniques

Given a graph that could, for example, represent a computer network connectivity scheme, there are different search techniques to find optimal values to get from one point of the graph to another. The following section is a discussion of how some of these search techniques work, more specifically, the breadth first search, depth first search, and best first search techniques.

1.5.1 Breadth First Search

This method (Figure 1.4) consists of expanding the search through the neighboring nodes. In this case, the algorithm begins searching all the directly connected nodes. The nodes through which it passes to reach the goal node are stored in a queue, and once the last connected node is reached, it starts removing from the queue and continues executing the algorithm, expanding successively.

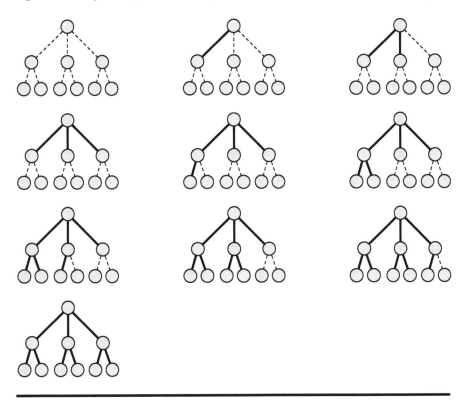

Figure 1.4   The different steps of a tree search using the Breadth First Search method.

The following is an example of the Breadth First Search algorithm:

```
Algorithm BFS(G, s)
    L₀ ← NULL
    L₀ insert_Last(s)
    setLabel(s, VISITED)
    i ← 0
    while ¬Lᵢ.isEmpty()
        Lᵢ ₊₁ ← NULL
        for all v ∈ Lᵢ.elements()
            for all e ∈ G.incidentEdges(v)
                if getLabel(e) = UNEXPLORED
                    w ← opposite(v,e)
                    if getLabel(w) = UNEXPLORED
                        setLabel(e, DISCOVERY)
                        setLabel(w, VISITED)
                        Lᵢ ₊₁.insertLast(w)
                else
                        setLabel(e, CROSS)
        i ← i +1
end BFS
```

1.5.2 Depth First Search

This method (Figure 1.5) consists of expanding the search for the deepest nonterminal node. This algorithm begins its search on a tree branch until a goal node is reached. The nodes through which it passes to reach the goal node are stored in a queue, and once it reaches the goal node, it starts removing from the queue and continues executing the algorithm, expanding in depth successively.
The Depth First Search algorithm is as follows:

```
Algorithm DFS(G, v)
    Input graph G and a start vertex v of G
    Output labeling of the edges of G
    in the connected component of v
```

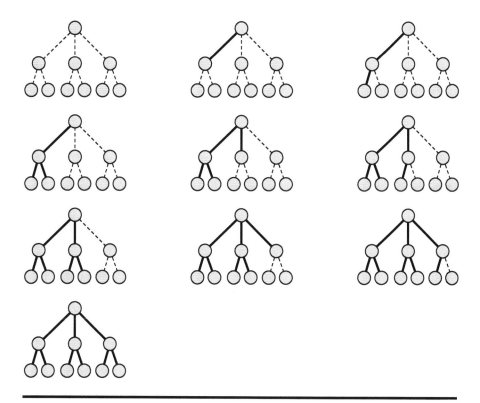

Figure 1.5 The different steps of searching a tree using the Depth First Search method.

```
        as discovery edges and back edges
        setLabel(v, VISITED)
        for all e ← G.incidentEdges(v)
            if getLabel(e) = UNEXPLORED
                w ← opposite(v,e)
                if getLabel(w) = UNEXPLORED
                    setLabel(e, DISCOVERY)
                    DFS(G, w)
                else
                    setLabel(e, BACK)
end DFS
```

1.5.3 Best First Search

This technique (Figure 1.6) expands the search of the most promising node; it can be considered an improved Breadth First Search. In this case, each tree branch will have an associated weight, and the decision on which direction to take in the search will be based on the value of such weight.

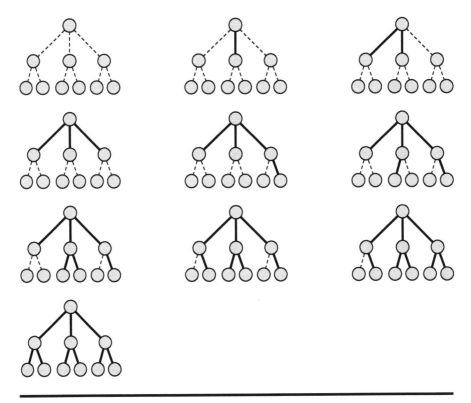

Figure 1.6   The different steps of a tree search using the Best First Search.

Chapter 2

Multi-Objective Optimization Concepts

Optimization problems are normally stated in a single-objective way. In other words, the process must optimize a single-objective function complying with a series of constraints that are based on constraints given by the real world. A single-objective optimization problem may be stated as follows:

Optimize [minimize/maximize]

$$f(X) \qquad (1)$$

subject to

$$H(X) = 0$$

$$G(X) \leq 0$$

In this case, the function to be optimized (minimize or maximize) is $f(X)$, where vector X is the set of independent variables. Functions $H(X)$ and $G(X)$ are the constraints of the model.

For this problem we can define three sets of solutions:

1. The **universal set**, which in this case is all possible values of X, whether feasible or nonfeasible.

2. The **set of feasible solutions**, which are all the values of X that comply with the $H(X)$ and $G(X)$ constraints. In the real world, these variables would be all possible solutions that can be performed.

3. The **set of optimal solutions**, which are those values of X that, in addition to being feasible, comply with the optimal value (minimum or maximum) of function $f(X)$, whether in a specific $[a, b]$ interval or in a global context $(-\infty, \infty)$. In this case, one says that the set of optimal solutions may consist of a single element or several elements, provided that the following characteristic is met: $f(x) = f(x')$, where $x \neq x'$. In this case, we can say that there are two optimal values to the problem when vector $X = \{x, x'\}$.

But in real life, it is possible that when wanting to solve a problem, we may need to optimize more than one objective function. When this happens, we speak of multi-objective optimization.

Figure 2.1 illustrates the topology of a network with five nodes and five links, each with a specific cost, on which one wishes to transmit a File Transfer Protocol (FTP) package from node 1 to node 5. The problem consists of determining which of the two possible paths one must take to transmit this package. In Figure 2.1, a cost has assigned between each pair of nodes (link).

An analysis of Figure 2.1 shows that if we take the path given by links (1, 2) and (2, 5), we would have a 2 jump path and a cost of 20 units. If we take the path given by links (1, 3), (3, 4) and (4, 5), however, we would have a 3 jump path and a cost of 15 units. In this specific case, we can identify 2 feasible paths: the first one given by links (1, 2) and (2, 5) and the second given by links (1, 3), (3, 4) and (4, 5). If we only want to optimize (in this case minimize) the number of jumps, we can see that the minimum value is 2 jumps, and it is given by path (1, 2) and (2, 5). On the other hand, if we only want to minimize the cost objective function, the minimum value would be 15 units and would be given by path (1, 3), (3, 4) and (4, 5). This example shows that the optimal solution

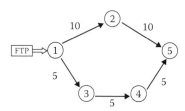

Figure 2.1   Network topology.

and value obtained depend on the objective function optimized, but with the drawback that other results will only be contemplated as feasible solutions, and not as optimal solutions.

In the previous example mentioned, if we optimize the number of jumps function, we will be paying the highest cost of transporting the package on this network. But it can also happen that the network administrator wants to use the path with fewer jumps. For this reason, we will state a model with more than one objective function, in which the solution will consist of a set of optimal solutions, not a single optimal solution.

A multi-objective optimization model may be stated as follows:

Optimize [minimize/maximize]

$$F(X) = \{f_1(X), f_2(X), \ldots, f_n(X)\} \tag{2}$$

subject to

$$H(X) = 0$$

$$G(X) \geq 0$$

In this case, the functions to be optimized (whether minimize or maximize) are the set of functions $F(X)$, where the vector X is the set of independent variables. Functions $H(X)$ and $G(X)$ are the constraints of the model. The solutions found solve the objective solutions even when they are conflicting, that is, when minimizing one function may worsen other functions.

Stating and solving multi-objective optimization problems applied to computer networks is the purpose of this book.

The first part of this chapter does a comparative analysis between performing single-function and multiple-function optimization processes. Next, the chapter shows some of the traditional methods used to solve multi-objective optimization problems, and finally, the last section shows some metaheuristics to solve multi-objective optimization processes.

2.1 Single-Objective versus Multi-Objective Optimization

When the optimization process is performed on a single-objective problem, only one function will be minimized or maximized, and therefore, it will be necessary to find a minimum or maximum, whether local or global, for that objective function. When we speak of multiple objective functions, we wish to find the set of values that minimize or maximize each of these functions. Moreover, a typical feature of this type of problem is that these functions may conflict with each other; in other words, when a function is optimized, other functions may worsen. In multi-objective optimizations we may find the following situations:

> Minimize all objective functions.
> Maximize all objective functions.
> Minimize some and maximize others.

The following example, which serves as an introduction to the formal theory of multi-objective optimization, explains the difference between optimizing a single function and optimizing multiple functions simultaneously.

Suppose you want to go from one city to another and there are different means of transportation to choose from: airlines, with or without a nonstop flight, train, automobile, etc. The actual duration and cost of the trip will vary according to the option selected.

Table 2.1 and Figure 2.2 show cost and trip duration data for each, independent of the means of transportation used.

An analysis of Table 2.1 or Figure 2.2 shows that by minimizing the time function only, we will reach the city of destination in 1 h, but at the highest cost ($600). By minimizing the cost function only, we will obtain the most inexpensive solution ($100), but the duration of the trip will be the longest (16 h).

By optimizing a single function, we obtain a partial view of the results, when in the real problem it could happen that more than one function influences the decision of which option to select.

However, if the problem is analyzed in a multi-objective context, the idea of this optimization process is that the result is shown as a set of optimal solutions (optimal Pareto front) for the different objective functions jointly. This can be seen in the table. Later on, we will show that all of these points are optimal in the multi-objective context. Therefore, if we were to optimize both functions simultaneously, these would be the values obtained.

Table 2.1 Multi-Objective Example

Option	1	2	3	4	5	6	7	8	9	10
Cost ($)	600	550	500	450	400	300	250	200	150	100
Time (hours)	1	2	3	4	5	6	8	10	14	16

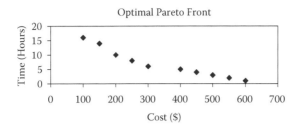

Figure 2.2   Optimal values in multi-objective optimization.

2.2 Traditional Methods

Solving the previous scheme can be quite complex. It requires optimizers that instead of providing a single optimal solution, yield as a result a set of optimal solutions, and such optimizers do not exist. In this section we will introduce some methods to solve multi-objective optimization problems in a single-objective context. This way, we can use the existing traditional optimizers. We will also show their drawbacks and constraints when finding the multiple solutions.

There are several methods that can be used to solve multi-objective problems using single-objective approximations. Some of them are weighted sum, ε-constraint, weighted metrics, Benson, lexicographic, and min-max, among others. All of these methods try to find the optimal Pareto front using different approximation techniques.

Later on we will analyze methods that use metaheuristics to solve multi-objective optimization problems in a real multi-objective context.

2.2.1 Weighted Sum

This method consists of creating a single-objective model by weighing the n objective functions by assigning a weight to each the functions.

Through the weighted sum method, the multi-objective model (2) can be restated in the following way:

Optimize [minimize/maximize]

$$F^{'}(X) = \sum_{i=1}^{n} r_i * f_i(X) \qquad (3)$$

subject to

$$H(X) = 0$$

$$G(X) \geq 0$$

$$0 \leq r_i \leq 1, \, i = \{1, \, ..., \, n\}$$

$$\sum_{i=1}^{n} r_i = 1$$

In this case one can see that each function is multiplied by a weight (w_i) whose value must be found between 0 and 1. Also, the sum of all the weights applied to the function must be 1.

An analysis of this type of solution shows that function $F(X)$ obtained from the sum is really a linear combination of the functions $f_i(X)$. What the weighted sum method does exactly is find the points of the optimal Pareto front, which consists of all the optimal solutions in the multi-objective optimization, through the combinations given by the weighted vector $R = \{r_1, \, r_2, \, ..., \, r_n\}$.

For example, in the case of two objective functions we would have the following equation:

$$F(X) = r_1 \cdot f_1(X) + r_2 \cdot f_2(X)$$

The idea is to assign values to r_1 and r_2 to find two lines that tangentially touch the set of solutions (K). If in Figure 2.3 one is minimizing functions

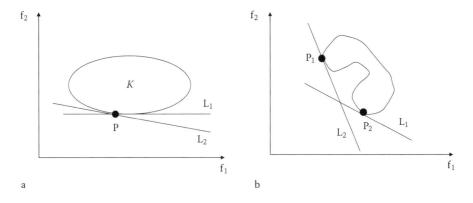

a   b

Figure 2.3 Solution using the weighted sum method.

F_1 and F_2 through weights (r_1, r_2), by means of this method the optimization process will find the crossing point P of the the two lines (linear combinations) L_1 and L_2. This point (P) would be the optimal solution found by means of this method with weights (r_1, r_2) and that belongs to the optimal Pareto front in the multi-objective context. If we were to change the values of (r_1, r_2), we would find another point P that would also belong to the optimal Pareto front.

If the set of solutions K is convex, any element of the optimal Pareto front may be found by changing the values of weight vector R (r_1 and r_2) in the example in Figure 2.3a. But this is only true when the set of solutions K is convex. If solution set K is nonconvex (Figure 2.3b), there are points of K that cannot be represented by a linear combination, and therefore, such points cannot be found by means of this method.

For this method to work completely, the search space for the set of solutions must be convex; otherwise, we cannot guarantee that all optimal Pareto front points will be found. In a convex set of solutions, all elements of the optimal Pareto front would have a $p > 0$ probability of being found, but in nonconvex sets of solutions, this probability could be 0 for certain elements of the optimal Pareto front.

Another drawback of this method deals with the range of values that can be obtained in the functions optimized. The problem arises when the ranges of the functions ($a \leq f_1 \leq b$, $c \leq f_2 \leq d$) have different magnitudes, and especially when such differences are very large. If this happens, the function having the highest range values will predominate in the result. For example, if $0 < f_1 < 1$ and $0 < f_2 < 1000$, all solutions would be biased to function f_2, which is the function with the larger-sized range.

To solve the previous drawback, it would be necessary to normalize the values of all objective functions, and this would generate a greater effort to obtain the results of the optimization model.

Example

The following shows how we can apply this method to solve a simple problem with two objective functions that we want to minimize:

$$f_1(x) = x^4$$

$$f_2(x) = (x - 2)^4$$

subject to

$$-4 \leq x \leq 4$$

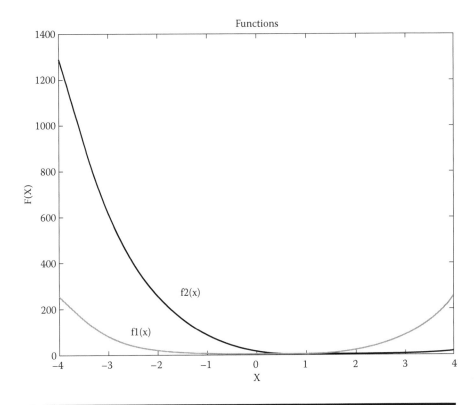

Figure 2.4 $f_1(x) = x^4$ and $f_2(x) = (x - 2)^4$ functions.

Figure 2.4 shows both functions in terms of independent variable x.

Because both functions in this case are convex, the weighted sum method can be used. According to this method, the optimization function would look as follows:

$$F(x) = w_1 \cdot f_1(x) + w_2 \cdot f_2(x)$$

subject to

$$-4 \leq x \leq 4$$

If we execute the model using any numerical solver for different values of w_i, we can obtain the following solutions for functions f_1 and f_2 (Table 2.2).

Table 2.2   Optimal Values with Weighted Sum

Sol	w_1	w_2	f_1	f_2	x
1	0	1	16.0	0	2.0
2	0.1	0.9	3.32	0.18	1.35
3	0.2	0.8	2.27	0.36	1.23
4	0.3	0.7	1.69	0.55	1.14
5	0.4	0.6	1.30	0.76	1.07
6	0.5	0.5	1	1	1
7	0.6	0.4	0.76	1.30	0.93
8	0.7	0.3	0.55	1.69	0.86
9	0.8	0.2	0.36	2.27	0.77
10	0.9	0.1	0.18	3.33	0.649
11	1	0	0	16	0

Figure 2.5 shows a graph with the optimal points obtained using the weighted sum method for the stated functions. The horizontal axis is associated with f_1, and the vertical axis is associated with function f_2. The figure only shows the values of f_1 and f_2 comprised between [0, 5]. For this reason, solutions 1 and 11 have not been shown.

2.2.2 ε-Constraint

This method consists of creating a single-objective model in which only one of the functions will be optimized and the remaining function not optimized becomes constraints in the model. The multi-objective optimi-

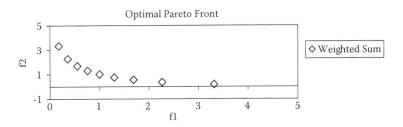

Figure 2.5   Optimal values with weighted sum.

zation model shown above (2) can be restated through the ε-constraint model as follows:

Optimize [minimize/maximize]

$$f_i(X) \tag{4}$$

subject to

$$f_k(X) \le \varepsilon_k, \; k = 1, \, ..., \, n \text{ and } k \ne i$$

$$H(X) = 0$$

$$G(X) \le 0$$

In this case, function $f_i(X)$ is the only one optimized; the other $n - 1$ functions become constraints and are limited by their corresponding values.

The objective in this method consists of changing the values of ε of each of the functions and, in this way, obtaining different optimization values in function $f_i(X)$.

Figure 2.6 shows an example in which two functions (f_1 and f_2) are optimized using the ε-constraint model. In the example, f_2 is optimized and f_1 has become a constraint.

For the two solutions P_1^a and P_1^b shown in Figure 2.6, one observes that when function f_1 is a constraint with upper limit ε_1^a, the optimization point found for f_2 is P_1^a. When in the model the constraint value for f_1 is changed to ε_1^b, the optimization value found for f_2 is P_1^b.

If we change the constraint value for f_1, we can obtain different values for f_2, and therefore obtain the values of the optimal Pareto front.

One can also optimize the other functions by changing the objective function to be optimized. In this case, the model would be

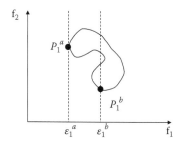

Figure 2.6   Solution using the ε-constraint method.

Optimize [minimize/maximize]

$$f_j(X) \tag{5}$$

subject to

$$f_k(X) \leq \varepsilon_k, \; k = 1, \; ..., \; n \text{ and } k \neq j$$

$$H(X) = 0$$

$$G(X) \leq 0$$

Here, function $f_i(X)$ is now a constraint and function $f_j(X)$ is the objective function. In this case, if the values of the functions that act as constraint are modified, one can find new values of the optimal Pareto front.

Through this method one can solve problems with convex and non-convex solution spaces. The drawback is that one must know the problem and the ε values for the results obtained in the optimization to be true solutions to the specific problem.

Now we will show how this method can be applied to solve a simple problem with two objective functions. As in the previous section, we need to minimize both functions.

Example

Apply the ε-constraint method to the two previous functions.

Considering f_2 as the objective function and f_1 as the constraint, the model would be

$$\text{Min } f_2(x) = (x - 2)^4$$

$$\text{Min } f_2(x) = (x - 2)^4$$

$$\text{Subject to}$$

$$x^4 \leq \varepsilon_1$$

$$-4 \leq x \leq 4$$

In this case, given different values of ε_1, we can obtain the following solutions, again using any numerical solver, such as sparse non-linear optimizer (SNOPT). (Table 2.3)

Table 2.3   Optimal Values with ε-Constraint

Sol	ε_1	f_1	f_2	X
1	20.0	16	0	2
2	5.0	5	0.06	1.50
3	4.5	4.5	0.09	1.46
4	3.0	3.0	0.22	1.32
5	2.5	2.5	0.3	1.26
6	2.0	2.0	0.43	1.19
7	1.5	1.5	0.64	1.11
8	1.0	1.0	1.0	1.0
9	0.5	0.5	1.81	0.84
10	0.3	0.3	2.52	0.74
11	0.2	0.2	3.14	0.67
12	0	0	15.67	0.01

Figure 2.7 shows the optimal points obtained using the weighted sum method and the ε-constraint method for the functions discussed.

2.2.3 Distance to a Referent Objective Method

This method, like the weighted sum method, allows one to transform a multi-objective optimization problem into a single-objective problem. The function traditionally used in this method is distance.

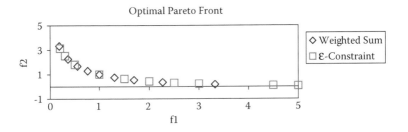

Figure 2.7   Optimal values with ε-constraint versus weighted sum.

Through the distance to a referent objective method, the multi-objective method (2) can be rewritten as follows:

Optimize [minimize/maximize]

$$F'(X) = \left[\sum_{i=1}^{n} \left| Z_i - f_i(X) \right|^r \right]^{1/r}$$
(6)

subject to

$$H(X) = 0$$

$$G(X) \geq 0$$

$$1 \leq r < \infty$$

In this method, one must set a Z_i value for each of the functions. These Z_i values serve as referent points to find the values of the optimal Pareto front by means of the distance function, as shown in Figure 2.8. Another value that must be set is r, which will tell us the distance function to be used. For example, if $r = 1$ is a solution equivalent to the weighted sum method but without normalizing the values of the functions; if $r = 2$ one would be using the Euclidean distance and if $r \rightarrow \infty$ the problem is called the min-max or Tchebychev problem.

In the specific case of $r \rightarrow \infty$, the formula of the method can be written as

Optimize [minimize/maximize]

$$F'(X) = \max_{i=1}^{n} \left[\left| Z_i - f_i(X) \right| \right]$$
(7)

subject to

$$H(X) = 0$$

$$G(X) \leq 0$$

$$1 \leq r < \infty$$

Figure 2.8 shows examples of this method with two functions and using r with values 1, 2, and ∞.

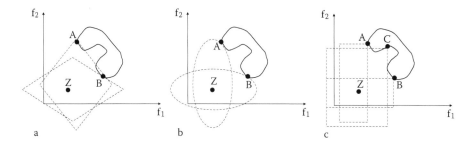

Figure 2.8 Solution using the distance to a referent objective method.

As can be seen in Figure 2.8, this method is sensitive to the values of Z_i and r. In Figure 2.8a and 2.8b, one can see that through this method it is not possible to find certain values of the optimal Pareto front between A and B. However, through Figure 2.8c (with r ➔ ∞), it is possible to find such points.

It may not be possible to find solutions in the optimal Pareto front for certain values of r when the set of solutions is nonconvex. Therefore, if the solution space is nonconvex or one is uncertain of it being convex, one should use the method with r ➔ ∞. Finally, analyzing the graphs in Figure 2.8, one can see that it is necessary to provide a good value of Z as point of reference. Providing a poor point Z may result in divergence instead of convergence toward optimal. This can be a problem due to ignorance of the set of solutions.

2.2.4 Weighted Metrics

This method is similar to the previous one, with the difference that each of the functions is normalized with respect to a weight. Using the weighted metrics method, the multi-objective model (2) can be rewritten by adding a weight-multiplying factor to each function to normalize their values.

Optimize [minimize/maximize]

$$F'(X) = \left[\sum_{i=1}^{n} w_i \left| Z_i - f_i(X) \right|^r \right]^{1/r} \tag{8}$$

Subject to

$$H(X) = 0$$

$$G(X) \leq 0$$

$$1 \leq r < \infty$$

$$0 \leq w_i \leq 1, \; i = \{1, \, ..., \, n\}$$

$$\sum_{i=1}^{n} w_i = 1$$

In the specific case of $r \rightarrow \infty$, the formula of the method can be rewritten in the following way:

Optimize [minimize/maximize]

$$F'(X) = \max_{i=1}^{n} \left[w_i \left| Z_i - f_i(X) \right| \right] \tag{9}$$

Subject to

$$H(X) = 0$$

$$G(X) \geq 0$$

$$1 \leq r < \infty$$

$$0 \leq w_i \leq 1, \; i = \{1, \, ..., \, n\}$$

$$\sum_{i=1}^{n} w_i = 1$$

This method has the same drawbacks as the method previously mentioned.

Example

Apply this method when $r \rightarrow \infty$ to the two previous objective functions:

$$f_1(x) = x^4 \; y \; f_2(x) = (x - 2)^4$$

$$\text{Min } z = \max \left[w_1 \cdot \left| z_1 - f_1(x) \right|, \; w_2 \cdot \left| z_2 - f_2(x) \right| \right]$$

Subject to

$$-4 \leq x \leq 4$$

If we execute the different values of w_i and z_i, we obtain the following solutions: (Table 2.4)

Table 2.4 Optimal Values with Weighted Metrics

Sol	w_1	w_2	z_1	z_2	f_1	f_2	x
1	0	1	0	0	15.87	0	2.0
2	0.1	0.9	0	0	2.59	0.29	1.27
3	0.2	0.8	0	0	1.88	0.47	1.17
4	0.3	0.7	0	0	1.49	0.64	1.11
5	0.4	0.6	0	0	1.22	0.81	1.05
6	0.5	0.5	0	0	1	1	1
7	0.6	0.4	0	0	0.81	1.22	0.95
8	0.7	0.3	0	0	0.64	1.49	0.89
9	0.8	0.2	0	0	0.47	1.88	0.83
10	0.9	0.1	0	0	0.29	2.59	0.73
11	1	0	0	0	0	16	0

Figure 2.9 shows the optimal points obtained using the weighted metrics method for the given functions and compares them with the methods previously analyzed for this same example.

2.2.5 The Benson Method

This method is similar to the weighted metrics method, but in this case the Z reference value has to be a feasible solution. Using the Benson method, one can rewrite the multiobjective model (2) the following way:

Maximize

$$F'(X) = \sum_{i=1}^{n} \max\left(0, Z_i - f_i(X)\right)$$

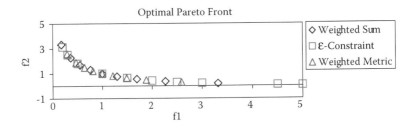

Optimal Pareto Front

Figure 2.9 Optimal values with weighted metric versus ε-constraint versus weighted sum.

Subject to

$$f_i(X) \leq Z_i, \ i = 1, 2, \ldots, n$$

$$H(X) = 0$$

$$G(X) \leq 0$$

As can be seen, instead of minimizing, one is maximizing the resulting function F because this method works this way. In this case, objective functions will be constraints by vector Z. Using this method, one can obtain the Pareto front values in a nonconvex solution space. The drawback consists of finding good values for Z to find the points in the Pareto front. One must know the space of feasible solutions of the problem being solved or, in lack of it, generate such values randomly.

Example

Apply this method to the two objective functions already mentioned:

$$\text{Min } z = \max [0, z_1 - f_1(x)] + [0, z_2 - f_2(x)]$$

Subject to

$$x^4 \leq z_1$$

$$(x - 2)^4 \leq z_2$$

$$-4 \leq x \leq 4$$

If we execute the method for different values of w_i and z_i, we obtain the following solutions (Table 2.5):

Table 2.5 Optimal Values with the Benson Method

Sol	z_1	z_2	f_1	f_2	x
1	5	2	0.43	2	0.81
2	5	1	1	1	1
3	3	4	0.19	4	0.59
4	2	5	0.06	5	0.51
5	1	3	0.22	3	0.68

Figure 2.10 shows the optimal points obtained using the weighted metrics method for the given functions and compares them with the methods previously analyzed for this same example.

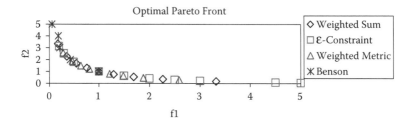

Figure 2.10   Optimal Benson values versus other methods.

After analyzing the different methods and applying them to a simple case, we can say that these methods converge toward the optimal Pareto front, although one must take into consideration the specific drawbacks of each.

There are other traditional methods that can be used to solve these multi-objective problems. They include the Keeney–Raiffa, compromise, goal attainment, lexicographic, proper equality constraints, and goal programming methods and others that belong to the group of interactive methods and will not be considered in this book.

2.3 Metaheuristics

Metaheuristics are considered high-level algorithmic strategies that are used to guide other heuristics or algorithms in their search of the space of feasible solutions of the optimal value (in the single-objective case) and the set of optimal values (in the multi-objective case).

Traditionally, metaheuristics are good techniques to solve optimization problems in which to accomplish convergence toward optimum one must perform combinatorial analysis of solutions. The way to find some new solutions that are unknown thus far and that belong to the set of feasible solutions is done through the solutions previously known. These techniques accomplish the following functional features.

2.3.1 Convergence toward Optimal

As the algorithm is executed, feasible solutions found must converge toward the optimal value.

Figure 2.11 shows that when both functions (f_1 and f_2) are minimized, the solutions found in iteration $i + 1$, compared to solutions found in iteration i, converge to the optimal Pareto front. This does not mean that every new solution obtained has to be better than the previous one, but this process does have to show this behavior.

2.3.2 Optimal Solutions Not Withstanding Convexity of the Problem

When metaheuristics function through combinatory analysis, any solution belonging to the set of feasible solutions must have a probability $p > 0$ of being found. If a metaheuristic meets this requirement, then any optimal solution may be found. In this case, it does not matter whether the set of solutions is convex, because this technique will always find the optimal solution.

Figure 2.12 shows that despite having a nonconvex set of solutions, the solutions of iteration $i + 1$ can be found through combinatorial analysis of the solutions obtained in iteration i.

2.3.3 Avoiding Local Optimal

There are heuristic algorithms that can find local optimal. If the problem consists of finding local optima, one does not need to use metaheuristic techniques to solve it. If the problem consists of finding global optimal, such metaheuristic techniques are required to not fall in local optimal; in

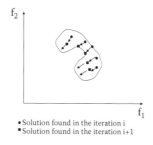

• Solution found in the iteration i
▪ Solution found in the iteration i+1

Figure 2.11   Convergence toward optimal.

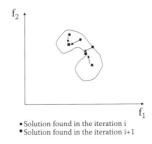

Figure 2.12   Solution in nonconvexity.

other words, such local optima can be found, but the final solution must be the global optimal value. This can be accomplished through combinatory analysis functions. Instead of dismissing a poor value found, it is left as a solution, and perhaps, through this poor value, one may find a global optimum.

Figure 2.13 shows that even though the solution found in iteration *i* + 1 is worse than the solutions found in iteration *i*, this solution is drawing the optimization process from such local optimum and is drifting it toward another optimum (which in the example of the figure corresponds to the global optima). The solution found in iteration *i* + 2 is better than the one found in iteration *i* + 1, but this latter one was worse than the one found in iteration *i*. Metaheuristics usually implement this functioning precisely to leave the local optima and be able to converge toward global optima.

2.3.4 Polynomial Complexity of Metaheuristics

This is one of the best contributions that these techniques can make to the solution of engineering problems; it consists of metaheuristics being used when, due to the complexity of the problem, one has not found

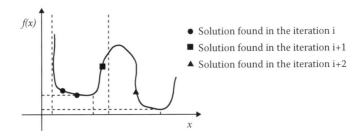

Figure 2.13   Global optimum.

algorithmic solutions with polynomial computational time. Because of their probabilistic work scheme, metaheuristics traditionally solve the problem in polynomial time. This, for example, implies that even though we cannot guarantee that the optimal value will necessarily be found, we can find a good approximate value that we can actually use. By being a probabilistic process, it can also happen that one finds a value in one execution and a different, but equally good value, in another execution.

The foregoing features are among the most important exhibited by these types of techniques to solve optimization problems. Different metaheuristics have been developed through time. Some have been based on biological processes, others on engineering processes, and others on social or cultural processes, etc. Some of the metaheuristics developed so far include evolutionary algorithms, ant colony, memetic algorithm, Tabu search, and simulated annealing. Other techniques used to solve optimization problems include scatter search, cultural algorithms, neuron networks, etc., but they will not be covered in this book.

2.3.5 Evolutionary Algorithms

Evolutionary algorithms are among the types of algorithms that solve combinatorial optimization problems and whose traditional solutions are very complex at a computational level.

Evolutionary algorithms originate in Darwin's theory of evolution, which explains the creation of species based on the evolution of life on Earth. Darwin introduced three fundamental components of evolution: replication, variation, and natural selection. Replication is the formation of a new organism from a previous one, but replicating would only produce identical copies of organisms, thereby stalling evolution. However, during the replication process there occur a series of errors called variations that allow a change in individuals. One manner of variation is sexual reproduction. In addition to replication and variation, evolution needs natural selection, which happens when individuals of a same species compete for the scarce resources of their environment and the possibility of reproducing. Such competition allows for the fittest individuals to survive and the weakest to die.

Simulating the biological evolutionary process, evolutionary algorithms use a structure or individual to solve a problem. This representation is generally a bit chain or data structure that corresponds to the biological genotype. The genotype defines an individual organism that is expressed in a phenotype and is made up of one or several chromosomes, which in turn are made up of separate genes that take on certain values (alleles) of a genetic alphabet. A locus identifies a gene's position in the chromosome. Hence, each individual codifies a set of parameters used as entry of the function under consideration. Finally, a set of chromosomes is called the population. Just as in nature, varied evolutionary operators function in the algorithm population trying to produce better individuals. The three operators most used in evolutionary algorithms are mutation, recombination, and selection.

2.3.5.1 Components of a General Evolutionary Algorithm

An evolutionary algorithm consists of the individuals, the population of such individuals, and the fitness of each of the individuals and of the genetic operators. We will explain in a very simple manner what each of these components consists of.

2.3.5.1.1 Individual

As previously mentioned, an individual is the representation of a solution to the problem. It includes bit chains, trees, graphs, or any data structure adjusted to the problem being solved.

2.3.5.1.2 Population

A population is the set of individuals used by the algorithm to perform the evolutionary process. The initial population generally is generated randomly. Subsequently, it is modified iteratively by the operators with the purpose of improving the individuals.

2.3.5.1.3 Aptitude Function (Fitness)

The aptitude function assigns each individual a real value that shows how good such an individual's adaptation is to its environment. In many algorithms, especially those that optimize an objective, this fitness is equal to the function optimized. In multi-objective algorithms, however, it is a more elaborate function that takes into account other important factors in this type of problem, for example, how many individuals are close among them or how many individuals dominate a specific individual from another population.

2.3.5.1.4 Genetic Operators

Genetic operators are applied on individuals of the population, modifying them to obtain a better solution. The concept of population improvement is very dependent on the problem. In a multi-objective problem, as previously mentioned, it is important to obtain a Pareto front as close to the real one as possible. The three most common operators used in an evolutionary algorithm are selection, recombination, and mutation.

2.3.5.1.4.1 Selection Operator — To obtain the next population, genetic operators are applied to the best or fittest individuals (set of parents). The roulette and tournament methods are the two most common selection operators.

Roulette Method — Let Ω be the set of individuals that represent the feasible solutions to the problem and $f : \Omega \to \Re$ the fitness function. If $x \in \Omega$, then $p(x) = \dfrac{f(x)}{\sum\limits_{y \in \Omega} f(y)}$ defines the probability of individual x being part of the set of parents.

Tournament Method — Let Ω be the set of individuals representing the feasible solutions to the problem and $f : \Omega \to \Re$ the fitness function. The set of parents $S \subset \Omega$ will be made by elements $x \in S$ such that $f(x) = \max(f(y))_{y \in S}$. To select the elements that will make up the set of parents in this method, form small subsets (usually randomly).

2.3.5.1.4.2 Crossover Operator — Let Ω be the set of individuals representing the feasible solutions to the problem. A crossover operator is a function $\lambda : \Omega x \Omega \to \chi \rho(\Omega)$, where $\rho(\Omega)$ is the set of all subsets of Ω. The crossover function combines two elements of the current population to produce one or more offspring.

Figure 2.14 illustrates an example of the crossover operator, where, by selection of two chromosomes (parents), one produces two offspring chromosomes. It is possible that operators select more than two parents and also produce more than two offspring.

2.3.5.1.4.3 Mutation Operator — Let Ω be the set of individuals representing the feasible solutions to the problem. A mutation operator is a function $\beta : \Omega \to \Omega$ that takes an individual and transforms it into another.

Figure 2.15 illustrates an example of the mutation operator, where the selected chromosome is altered to produce a new chromosome. Mutation operators may alter more than one value in the chromosome.

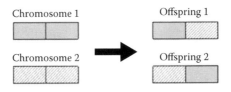

Figure 2.14   Crossover function.

Figure 2.15   Mutation function.

2.3.5.2 General Structure of an Evolutionary Algorithm

In this book we will show other methaheuristics that are also used to solve optimization problems, although greater emphasis will be placed on evolutionary algorithms.

Figure 2.16 is the general structure of an evolutionary algorithm to solve an optimization algorithm. Population P_0 initiates in line 2. These individuals are usually generated randomly.

Normally, the end condition of these algorithms is the performance of a constant number of iterations. In certain cases, additional conditions are introduced to control convergence of the algorithm to the optimal value sought.

In P one assigns the individuals that the selection operator takes from P_t (line 4). In line 5 one selects from P the individuals that will be the parents of the next generation, that is, upon which one will apply the crossover operators.

In line 6 one assigns a P' to all the individuals produced with the crossover operator. To a subset of such individuals one applies the mutation operator (line 7). In line 9 one determines all the individuals of the next generation (P_{t+1}). The individuals that continue into the next generation are usually the individuals produced with the crossover and mutation operators. However, certain algorithms maintain high-quality individuals from previous generations.

Evolutionary algorithms comply with the following metaheuristic characteristics:

t: iteration number.

P_t: Population in iteration t.

pc: Probability of performing crossover.

pm: Probability of performing mutation.

  : Crossover operator.

ß: Mutation operator.

S: is a subset of the merger P' and P' obtained randomly

W: is a subset of P'' obtained randomly

Selection Operator(): selection operator.

Begin:

1.   t=0

2.   Initialize P_0

3.   **While** (condition of completion $==$ false)

4.   P'=Selection_Operator(P_t)

5.   Select a Subset $S \subseteq P' x P'$

6. $P'' = \{\lambda(x) \mid \forall x \in S\}$

7.   Select a Subset $W \subseteq P''$

8. $P''' = \{\beta(x) \mid \forall x \in W\}$

9. $P_{t+1} = M \subseteq P_t \cup P'' \cup P'''$

10.   t=t+1

11.   EndWhile

End.

Figure 2.16 Evolutionary algorithm.

Convergence to optimum: In the multi-objective case, this usually takes place through the elitist or dominant functions. This way, one can eliminate the poor individuals in each generation, while only the good individuals remain. Because the process is iterative, one expects each generation to present better individuals, guaranteeing convergence toward optimal.

Optimal solutions not withstanding convexity of the problem: This feature is accomplished by means of the combinatory analysis produced through the crossover and mutation operators.

Avoid local optima: This feature is accomplished through the values of crossover and mutation probabilities. If the mutation probability is high, the algorithm may converge toward local optima. Normally, one recommends a value of 0.05 or 0.1 as a maximum. Another fact that can help accomplish this feature is that there may be nonoptimal individuals in the population, confirming what is illustrated in Figure 2.8 — that operators can produce a good individual from bad individuals.

Polynomial complexity of metaheuristics: Being a probabilistic algorithm, its consequence is that its execution is nonexhaustive. For more information about evolutionary algorithms, read [BAC00] and [DEB01].

2.3.6 Ant Colony

Optimization based on ant colonies (ACO) is a metaheuristic approach to solve combinatorial optimization problems. This metaheuristic is based on the foraging behavior of certain ant species. The ants cooperate among them to find food quickly. This cooperation is possible due to the indirect communication of the ants.

Initially, scouting ants start searching for food sources. As they return to the colony, they leave markers, called pheromones, on the path leading to the food, which will lead other ants on the right track. The amount of pheromones tells other ants how good a path is. When others follow the path, they leave their own markers. As time goes by and more ants take the path, the pheromones increase and, therefore, there is greater probability of the path being taken. In the absence of pheromones, ants randomly scout for paths.

To explain this behavior, Dorigo [DOR04] designed an experiment that served as a starting point for combinatorial optimization algorithms based on ant colonies. This experiment established two paths, one long and one short, between the colony and the food source. During these experiments he observed that ants tend to take the shorter path.

2.3.6.1 General Structure of an Ant Colony Algorithm

Figure 2.17 shows the pseudo-code of an algorithm based on ant colony. Three procedures must be implemented: construct ant solutions, update pheromones, and daemon actions. Dorigo et al. [DOR97] and [DOR04] say that this metaheuristic does not specify whether these procedures

Procedure ACOMetaheuristic

1.　　ScheduleActivities

2.　　　　ConstructAntsSolutions

3.　　　　UpdatePheromones

4.　　　　DaemonActions

5.　　End-ScheduleActivities

end-Procedure

Figure 2.17　Ant colony algorithm.

must be executed fully parallel and independently or some sort of synchronization is needed.

2.3.6.1.1 Ant-Based Solution Contruction

This is the procedure responsible for managing the ants that concurrently and asynchronically visit the adjacent states of the problem considered, moving through the neighbors of the construction graph. The transition from one state to the next is done using the pheromone trace and the heuristic information of how to recalculate the value of the pheromones. In this fashion, the ants construct the solutions that will later be evaluated with the objective, so that the pheromone update can decide what amount of pheromone to add.

2.3.6.1.2 Pheromone Trail Update

This is the procedure by which pheromone trails are modified. The pheromone trail in the link, in the case of application to computer networks, can increase by pheromone addition from passing ants or decrease due to pheromone evaporation. From a practical point of view, an increase in the amount of pheromone increases the possibility of an ant using a component or link to build the solution. On the other hand, pheromone evaporation is used to avoid too rapid convergence toward suboptimal regions in the space of solutions, and would therefore favor the exploration of new areas in the search space.

2.3.6.1.3 Daemon Actions

This procedure is used to implement centralized actions that cannot be performed by ants. For example, local searches and processes that take the best solutions and deposit additional pheromone on the links or components.

For more information about ant colonies, read [DOR04].

2.3.7 Memetic Algorithm

A memetic algorithm is a combination of an evolutionary algorithm and a local search. This combination is used to accomplish algorithms that comply mainly with the following two characteristics: flexibility to incorporate specific heuristics to the problem and specialized operators, and efficiency to produce acceptable solutions in less time.

Some authors see memetic algorithms as a simulation of cultural evolution. This simulation is conceived as the transmission, imitation, modification, and selection of the ideas, trends, methods, and all information that defines a culture. The minimum unit of cultural transmission is called memes, and it is the equivalent of genders in biological evolution. In the context of cultural evolution, memetic algorithms are defined as those that make an analogy of cultural evolution.

Because the general scheme is basically the same, generically memetic algorithms can be conceived as a hybrid of evolutionary algorithms and local search, and the simulation of cultural evolution.

2.3.7.1 Local Searches

Before introducing the general structure of a memetic algorithm, one must explain local searches.

A neighborhood function can be defined as follows: Let (X, f) (where X is the set of feasible solutions and f is the function to be optimized) be an instance of a combinatorial optimization problem. One neighborhood function is a mapping $\xi : X \to \rho(X)$ (where $\rho(X)$ is the set of all subsets of X), which defines for every solution $i \in X$ a set $\xi(i) \subseteq X$ of solutions that are in a certain sense close to i. The set $\xi(i)$ is the neighborhood of solutions of i, and every $j \in \xi(i)$ is a neighbor of i. One must assume that $i \in \xi(i)$ is met for $i \in X$.

Once the neighborhood function is defined, one says that a local search is a method that, starting from a solution, continually tries to find the best solutions, searching among the neighbors (obtained by the neighborhood function) of such solution.

t: number of the iteration.

P$_t$: Population in iteration t.

pc: probability of performing a crossover.

pm: probability of performing a mutation.

: crossover operator.

ß: mutation operator.

: local search.

Selection Operator(): this is the selection operator.

Local_Search();

Begin:

1.   t=0

2.   Initialize P$_0$

3.   **While**(Condition of Completion == false)

4.   P'=Selection_Operator(P$_t$)

5.   Select a Subset c

6. $P'' = \{\lambda(x) \mid \forall x \in S\}$

7. $L = \{\xi(x) \mid \forall x \in P''\}$

7.   Select a Subset $W \subseteq P''$

8. $P''' = \{\beta(x) \mid \forall x \in W\}$

9. $L' = \{\xi(x) \mid \forall x \in P'''\}$

9. $P_{t+1} = M \subseteq P_t \cup L \cup L'$

10.   t=t+1

11.   **EndWhile**

End.

Figure 2.18   Memetic algorithm.

2.3.7.2 General Structure of a Memetic Algorithm

Figure 2.18 is a general structure of a memetic algorithm. Operators used in memetic algorithms are the same operators used in evolutionary algorithms. Memetic algorithms have the same characteristics as evolutionary algorithms because they operate similarly, except for the local search added to find better solutions in less time.

For more information about memetic algorithms, read [MER00] and [MOS03].

2.3.8 Tabu Search

Tabu search (TS) is a metaheuristic procedure used for a local search heuristic algorithm to explore the space of solutions beyond the simple local optimum. It has been applied to a wide range of practical optimization applications, producing an accelerated growth of Tabu search algorithms in recent years. Using TS or TS hybrids with other heuristics or algorithmic procedures, one can establish new records to find better solutions in programming, sequencing, resource assignment, investment planning, telecommunications problems, etc.

Tabu search is based on the premise that to rate the solution to a problem as intelligent, it must incorporate adaptive memory and sensible exploration (responsive). Adaptive memory is the mechanism by which the Tabu search technique ensures that solutions already found in previous steps are not repeated. This mechanism stores solutions or steps taken to find such solutions in a temporary memory. The use of adaptive memory contrasts with "forgetful" designs, like the ones inspired on physics or biology metaphors; with "rigid memory" designs, such as those exemplified by branching and delimitation; and with its "cousins" in artificial intelligence (AI), which are normally not managed with this type of memory schemes.

The importance of sensible exploration in Tabu search, whether in a deterministic or probabilistic implementation, is based on the assumption that a poor strategic selection may produce more information than a good random selection.

In a memory-using system, a poor selection based on a strategy may provide useful clues on how to make beneficial modifications to the strategy.

The basis for a Tabu search (TS) can be described as follows. Given a function $f(x)$ to be optimized in a set X, TS starts as any local search, going iteratively from one point (solution) to another until meeting a given stopping criteria. Each $x \in X$ has an associated neighborhood (or vicinity) $N(x) \subseteq X$, and every solution $x' \in N(x)$ can be reached from x by means of an operation called movement.

TS exceeds local search using a modification strategy of $N(x)$ as the local search continues, replacing it by a neighborhood $N^*(x)$. As said in our previous discussion, a key aspect in Tabu search is the use of memory structures that help determine $N^*(x)$, and thus organize the way the space is explored.

The solutions admitted in $N^*(x)$ by these memory structures are determined in several ways. Specifically, the one that gives its name to Tabu search identifies solutions found on a specified horizon (and, implicitly, certain solutions identified with them) and excludes them from $N^*(x)$, classifying them as Tabu. In principle, the Tabu terminology implies a type of inhibition that incorporates a cultural connotation (i.e., something that is subject to the influence of history and context), which can be overcome under appropriate conditions.

The process by which solutions acquire a Tabu status has several facets designed to promote an aggressive examination guided by new points. One useful way to visualize and implement this process is to replace the original evaluation of solutions by Tabu evaluations, which introduce penalties to significantly discourage the election of Tabu solutions (i.e., that will preferably be excluded from the $N^*(x)$ neighborhood, according to their dependence with the elements that make up the Tabu status). Further, Tabu evaluations also periodically include incentives to stimulate the selection of other types of solutions as a result of aspiration levels and long-term influence.

TS profits from memory (and hence form the learning process) to perform these functions. Memory used in TS may be explicit or based on attributes, although both modes are not excluding. Explicit memory conserves complete solutions and typically consists of an elite list of solutions visited during the search (or highly attractive but unexplored neighborhoods for such solutions). These special solutions are strategically introduced to expand $N^*(x)$, and thus present useful options not found in $N(x)$.

The memory in TS is also designed to introduce a more subtle effect in the search by means of a memory based on attributes that saves information about solution attributes that change when moving from one solution to another. For example, in a graph or network context, attributes may consist of added nodes or arcs that are added, deleted, or replaced by the movements executed. In more complex problem formulations, attributes may represent function values. Sometimes attributes can also be strategically combined to create other attributes through hashing procedures or IA-related segmentations, or by vocabulary construction methods.

2.3.8.1 General Structure of a Tabu Search

Figure 2.19 is the general structure of an algorithm through Tabu search. This algorithm randomly finds an initial solution. Subsequently, and although the exit condition has not been met, one finds new solutions based on the existing one. These new solutions must be neighbors of the current solution and also cannot be in the Tabu list. The Tabu list is used

F represents feasible solutions

x' is the best solution found so far

c: iteration counter

T set of "tabu" movements

N(x) is the neighborhood function

1. Select x $\in$ F

2. x' = x

3. c = 0

4. T = ϕ

5. If {N(x) − T} = ϕ, goto step 2

6. Otherwise, c $\leftarrow$ c + 1

   Select n_c $\in$ {N(x) − T} such that: $n_c(x)$ = opt(n(x) : n $\in$ {N(x) − T)}

   opt() is an evaluation function defined by the user

7. x $\leftarrow$ $n_c(x)$

   If f(x) < f(x') then x' $\leftarrow$ x

8. Check stopping conditions:

   Maximum number of iterations has been reached

   N(x) − T = ϕ after reaching this

   step directly from step 2.

9. If stopping conditions are not met, update T

   and return to step 2

Figure 2.19   Tabu search algorithm.

to ensure that solutions that have been visited are not visited again. For more information about Tabu search, read [GLO97].

2.3.9 Simulated Annealing

Simulated annealing uses concepts originally described for statistical mechanics. According to Aarts [AAR1989], annealing is a thermal process in which, through a thermal bath, low energy status is produced in a solid. This physical process of annealing first softens the solid by heating it at a high temperature and then cools it down slowly until the particles rearrange themselves in a solid configuration. For each temperature reached in the annealing process, the solid can reach thermal balance if the heating occurs slowly. If cooling is done correctly, the solid reaches its fundamental state in which particles form perfect reticulate and the system is at its lowest energy level. However, if cooling takes place too rapidly, the solid may reach metastable status, in which there are defects in the form of high-energy structures.

The evolution of a solid in the thermal bath can be simulated using the Metropolis algorithm based on the Monte Carlos techniques. In this algorithm, thermal balance, described by the Boltzmann distribution, at a given temperature is obtained by generating an elevated number of transitions. Very briefly, the algorithm transitions from one step to the next according to the following rules: if the energy in the produced state is lower than the current state, then it accepts the produced state as the current state; however, the produced state will be accepted with a determined probability (based on the Boltzmann distribution). This acceptance probability is a function of temperature and the difference between the two energy levels. The lower the temperature, the lower the probability of transformation into a higher energy state, and the greater the energy in the new state, the lower the probability of it being accepted. Therefore, there is a possibility for every state to be reached, but with a different probability, depending on temperature.

Simulated annealing can be seen as an iterative Metropolis algorithms process that is executed with decreasing control parameter values (temperature). Conceptually, it is a search method by neighborhoods, in which the selection criteria are the rules of transition of the Metropolis algorithm: The algorithm randomly selects a candidate from the group that makes up the neighborhood of the current solution. If the candidate is better than it in terms of evaluation criteria, then it is accepted as the current solution; however, it will be accepted with a probability that decreases as the difference in cost between the candidate solution and the current solution increases. When a candidate is rejected, the algorithm randomly selects another candidate and the process is repeated.

Randomization in the selection of the next solution has been designed to reduce the probability of being trapped in a local optimum. One has been able to prove, as will be seen later, that simulated annealing is capable of finding (asymptotically) the optimal solution with one probability. Although guaranteed, optimality will only be reached after an infinite number of steps in the worst of cases. Therefore, asymptotic convergence of the algorithm can only be approximated in practice, which, fortunately, is done in polynomial time. Good functioning of the metaheuristic will depend largely on the design of the neighborhood structure, the cooling pattern, and the data structure.

In the early 1980s, working in the design of electronic circuits, Kirkpatrick et al. [KIR83] (and Cerny [CER85] doing independent work researching the Transmission Control Protocol (TCP)) considered applying the Metropolis algorithm to some of the combinatorial optimization minimization problems that appear in these types of designs. They felt it was possible to establish an analogy between the parameters in the thermodynamics simulation and the local optimization methods. Thus, they related the following:

Thermodynamic		Optimization
Configuration	⇔	Feasible solution
Fundamental configuration	⇔	Optimum solution
Energy of configuration	⇔	Cost of solution
Temperature	⇔	?

A real meaning in the field of optimization does not correspond to the physical concept of temperature; rather, it has to be considered a parameter, T, which will have to be adjusted. In a similar way, one could imagine the processes that take place when the molecules of a substance start moving in the different energy levels searching for a balance at a given temperature and those that occur in minimization (or maximization) processes in local optimization. In the first case, once the temperature is fixed, distribution of the particles in the different levels follows the Boltzmann distribution; therefore, when a molecule moves, this movement will be accepted in the simulation if the energy decreases, or will be accepted with a probability that is proportional to the Boltzmann factor in the opposite case. When speaking of optimization, once the T parameter is set, we produce an alteration and directly accept the new solution when its cost decreases. If the cost does not decrease, it is accepted with a probability that is proportional to the Boltzmann factor.

This is the key of simulated annealing, as it is basically a local search heuristic strategy where the selection of the new element in the neighborhood $N(s)$ is done randomly. As was seen previously, the drawback of this strategy is that if it falls in a local optimum during the search, it is unable to leave it. To avoid this, simulated annealing allows certain probability (lesser every time as we approach the optimal solution) passage to worst solutions. Indeed, analyzing the behavior of the Boltzmann factor as a function of temperature, we see that as temperature decreases, the probability of us accepting a solution that is worse than the previous one also decreases rapidly.

The strategy that we will follow in simulated annealing will start with a high temperature (with which we will allow changes to worse solutions during the first steps, when we are still far from the global optimum), and subsequently, we will decrease the temperature (reducing the possibility of changes to worse solutions once we are closer to the optimum sought). This procedure gives rise to the name of the algorithm: annealing is a metallurgical process (used, for example, to eliminate internal stress in cold laminated steel) through which the material is rapidly heated and then slowly cooled for hours in a controlled way.

2.3.9.1 General Structure of Simulated Annealing

Figure 2.20 shows the general structure of the algorithm through simulated annealing. Input parameters for this algorithm are the initial temperature, T_o, and the cooling velocity, α. This cooling velocity is associated with

Input (To, α, Tf)

1. T $\leftarrow$ To

2. Sact $\leftarrow$ Initial_Solution

3. While T $\geq$ T$_f$ do

4.   Begin

5.   For cont $\leftarrow$ 1 to L(T) do

6.   Begin

7.   Scand $\leftarrow$ Select_Solution_N(Sact)

8. $\delta \leftarrow$ cost(Scand) – cost(Sact)

9.   If $(U(0,1) < e^{(-\delta/T)})$ or $(\delta < 0)$ Then

10.   Sact $\leftarrow$ Scand

11.   End

12.   T $\leftarrow \alpha(T)$

13.   End

14.   Write the best solutions of Sact visited.

Figure 2.20   Simulated annealing algorithm.

the way we update the value T_{i+1} from T_i when the temperature decreases after $L(T)$ iterations.

An initial solution that belongs to the solution space Ω is generated and, until the end of the process is reached, for every T one calculates, a number $L(T)$ tells the number of iterations (before the temperature reduction) that a solution will combine. These new solutions found must be in the neighborhood $N(Sact)$ of the current one, which will replace the current solution if it costs less or has a probability $e^{(-\delta/T)}$. To calculate this probability, generate a uniform random number $[0, 1)$ that is represented as $U(0, 1)$. Finally, the solution offered will be the best of all the *Sact* visited. For more information about simulated annealing, read [VAN89].

2.4 Multi-Objective Solution Applying Metaheuristics

As we have seen in previous sections, the use of traditional techniques allows us to obtain optimal Pareto front points, yet some of these techniques have certain constraints. Something to bear in mind is that computer network problems are usually NP problems. Hence, to obtain an optimal solution one resorts to metaheuristics. Also bear in mind that the metaheuristics explained previously are aimed at providing a solution to single-objective problems, and therefore must be adjusted to solve multi-objective problems.

In the case of evolutionary algorithms, the Multi-Objective Evolutionary Algorithm (MOEA) is an adaptation to solve multi-objective problems. Similarly, other metaheuristic techniques can be adapted to address multi-objective problems. In this section we will discuss the MOEA technique.

There are different MOEA algorithms, which can be divided into elitist (SPEA, PAES, NSGA II, DPGA, MOMGA, etc.) and nonelitist (VEGA, VOES, MOGA, NSGA, NPGA, etc.). They can also be classified based on the generation in which they were developed. The first generation includes those that do not work with Pareto dominance (VEGA, among others) and those that do work with the Pareto dominance concept (MOGA, NSGA, NPGA, and NPGA 2, among others). The second generation includes PAES, PESA, PESA II, SPEA, SPEA 2, NSGA II, MOMGA, and MOMGA II, among others.

Because the purpose of this book is to show how metaheuristics solve computer network problems, we will not provide an analysis of the different evolutionary algorithms. One can consult some of these algorithms in books by [DEB01] and [COE02]. We will concentrate on the application of the SPEA algorithm to solve multi-objective problems through MOEA.

SPEA is a combination of new and previously used techniques to find different optimal Pareto and parallel solutions. It shares the following characteristics with other algorithms previously proposed: It saves the nondominated solutions it has found so far in an external population, thus implementing elitism. It uses the concept of Pareto dominance to assign a fitness value to individuals. It does clustering [MOR80] to reduce the number of nondominated solutions it has stored without destroying the characteristics of the Pareto front outlined so far.

On the other hand, SPEA exclusively uses the following concepts: It combines the above-mentioned three characteristics in a single algorithm (so far they had always been proposed separately). It determines fitness of an individual based solely on the solutions found in the population of nondominated individuals; this way it is not relevant that members of the general population dominate one another — all solutions are part of the external set of nondominant solutions that participate in the selection. It suggests a new method for inducing the formation of niches within the population, thus preserving genetic diversity; this method is based on the Pareto concept and does not require previous knowledge or determination of any distance parameter (as in the radius of the niche in SF (sharing function).

Figure 2.21 shows a pseudoalgorithm by means of SPEA.

2.4.1 Procedure to Assign Fitness to Individuals

One of SPEA's characteristics is the way it assigns fitness to each individual. The process in carried out so that the dominant and most characteristic individuals have better fitness values. At the same time, the process is thought to induce maintenance of genetic diversity and the obtaining of solutions distributed in the complete Pareto front.

1. Begin
2.   Generate randomly the initial population P_0 with size N
3.   Initialize the set P_E as an empty set
4.   Initialize the generation t counter to 0
5.   While $t < g_{max}$
6.   Evaluate the objectives on the members of P^t and $P_E{}^t$
7.   Calculate the fitness of each of the individuals in P^t and $P_E{}^t$
8.   Make the environmental selection to conform the new extern population $P_E{}^{t-1}$
9.   Apply the selection operator by binary tournament with replacement on $P_E{}^{t-1}$.
10.   Apply the crossover and mutation operators on the selected population.
11.   Assign the new population to P^{t-1}.
12. $t \leftarrow t+1$
13.   End While
14. End

Figure 2.21　SPEA algorithm.

The fitness assignment process takes place in two stages: First, a fitness value is assigned to the individuals in set P' (external, nondominated set). The members of P (set of general solutions) are then evaluated.

To assign fitness to members of the dominant set, each solution $i \in P'$ is assigned a real value $s_i \in [0,1)$, called strength, which is proportional to the number of individuals $j \in P$ for which $i \succ j$ (that is, it is a measure of the individuals dominated by i). Let n be the number of individuals in P that are dominated by i, and let us assume that N is the size of P. Then s_i is defined as $s_i = \dfrac{n}{N+1}$. The fitness of i is inversely equal to its strength: $F(i) = \dfrac{1}{s_i}$.

The strength of an individual $j \in P$ is calculated as the sum of strengths of all external, nondominated solutions $i \in P'$ that dominate j. Add 1 to that total to ensure that members of P' have better strength than the members of P. Hence, the fitness of individual j is stated as $f_j = \dfrac{1}{1 + \displaystyle\sum_{y \succ j} s_i}$,

where $s_j \in [1, N)$.

The idea that underlies this mechanism is to always prefer individuals that are closer to the optimal Pareto front and at the same time distribute them through the whole feasible surface.

2.4.2 Reducing the Nondominated Set Using Clustering

In certain problems, the optimal Pareto front can be very large and may even contain an infinite number of solutions. However, from the standpoint of the network administrator, once a reasonable limit is reached, no advantage is added by considering all the nondominated solutions found until that point.

Furthermore, the size of the external population of nondominated solutions alters the assignment procedure used by the algorithm. When the external dominant set exceeds the established limit, the algorithm no longer finds good solutions and also has greater probability of becoming stagnant in a local minimum. For these reasons, a purge of the external dominant set may not only be necessary, but even mandatory.

One method that has been applied successfully to this problem and has been studied extensively in the same context is the *cluster* analysis [MOR80]. Overall, the *cluster* analysis partitions a collection of p elements in q groups of homogenous elements, where $q < p$. Next, a characteristic or centroid element for the cluster is selected.

The average link method [MOR80] has been shown to be an adequate behavior in this problem and is the chosen one in [ZIT99] for the SPEA. For this, it is also the method used in the implementations subsequently described to control size of the external population of SPEA and prevent losing good and significant solutions already found.

The clustering procedure is expressed in the following pseudocode (Figure 2.22). In it we will show how SPEA can be applied to solve a simple problem with the same two objective functions worked with the traditional methods. As in the previous cases, we will minimize both functions but, in this case, in a real multi-objective context.

$$\text{Min } F(X) = [f_1(x) = x^4, f_2(x) = (x - 2)^4]$$

Subject to

$$-4 \leq x \leq 4$$

2.4.2.1 Representation of the Chromosome

In this case, one chromosome stands exactly for a value of x in the mathematical model. For this reason, it would be represented by a real value. For example,

Chromosome ➜ 2.5

As selection operator we have chosen the roulette method, by means of which we can obtain the chromosomes to execute the crossover and mutation operators.

2.4.2.2 Crossover Operator

By means of the roulette method we select, for example, the following chromosomes:

Chromosome 1 ➜ 2.5

Chromosome 2 ➜ 1.3

C: union of all clusters

i: one P' chromosome

P': nondominated population

P: dominated population

1. Clustering Procedure()

2. Begin

3. Initialize the set of *clusters* C; each individual $i \in P'$ constitutes a different *cluster*.

Hence: $C = \bigcup_i \{i\}$.

4. While $C > N'$

5.   Calculate the distance of all possible pairs of *clusters* where distance

$d = \dfrac{1}{|c_1| \cdot |c_2|} \cdot \sum_{i_1 \in c_1, i_2 \in c_2} \|i_1 - i_2\|$ of the two *clusters* c_1 and $c_2 \in C$ is the average distance

between pairs of individuals of these 2 *clusters*. Metrics $\| \|$ reflects the distance

between 2 individuals i_1 e i_2

6.   Determine two *clusters* c_1 and c_2 with minimum distance d and join them to

create a single one.

7. End While

8. Compute the reduced nondominated set, selecting an individual that is characteristic

of each *cluster*. The centroid (the point with minimum average distance with respect to

all other *cluster* points) is the characteristic solution.

9. End

Figure 2.22   Clustering procedure.

The crossover function takes the complete part of chromosome 1 (2) and the decimal part of chromosome 2 (.3), combining these two values to create the first offspring.

Offspring 1 ➔ 2.3

The second offspring is created taking the complete part of chromosome 2 (1) and the decimal part of chromosome 1 (.5).

Offspring 2 ➜ 1.5

2.4.2.3 Mutation Operator

In this case, we select one chromosome. For example,

Chromosome ➜ 2.5

The mutation function adds or subtracts a value δ. If a chromosome is selected for mutation, it randomly determines whether the value of δ is added or subtracted. Suppose that in this case we have $\delta = 0.01$ and one must add randomly. The result will be

Chromosome ➜ (2.5 + 0.01) = 2.51

Figure 2.23 shows the optimal points obtained by means of algorithm SPEA for the stated functions and compares them to the traditional methods previously analyzed in this same example. In this case, the following values for the SPEA have been used:

Probability of crossover ➜ 0.7

Probability of mutation ➜ 0.1

Maximum population size ➜ 50

Maximum number of generations ➜ 20

In this case, with a single execution we have obtained multiple values (represented by a filled-in circle) of the optimal Pareto front.

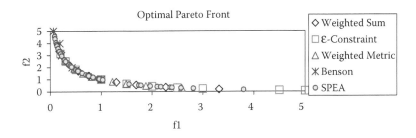

Figure 2.23 Optimal values of SPEA versus traditional methods.

Chapter 3

Computer Network Modeling

This chapter briefly introduces computer networks. Readers will learn basic concepts and fundamentals that will allow them to understand the application of metaheuristic techniques in multi-objective problems in computer networks.

3.1 Computer Networks: Introduction

A computer network can be defined as a set of computers that interact among the individual computers, sharing resources or information. Reference models have been developed in computer networks to establish the functions of certain layers in the network for proper performance. The following section introduces two reference models: the Open Systems Interconnection (OSI) model established by the International Organization for Standardization (ISO) and the Transmission Control Protocol/Internet Protocol (TCP/IP) model.

3.1.1 Reference Models

Designing, implementing, and manufacturing computer networks and related devices are very complex activities. Therefore, in order for this technology to be successful and massively used, the manufacturers' community saw the need to follow a series of standards and common models, allowing interoperability among the products of different brand names.

3.1.1.1 OSI Reference Model

The OSI reference model was developed by the ISO for international standardization of protocols used at various layers. The model uses well-defined descriptive layers that specify what happens at each stage of data processing during transmission. It is important to note that this model is not a network architecture because it does not specify the exact services and protocols that will be used in each layer.

The OSI model has seven layers. Each of the layers is described next.

3.1.1.1.1 Physical Layer

The physical layer is responsible for transmitting bits over a physical medium. It provides services at the data-link layer, receiving the blocks the latter generates when emitting messages or providing the bit chains when receiving information. At this layer it is important to define the type of physical medium to be used, the type of interfaces between the medium and the device, and the signaling scheme.

3.1.1.1.2 Data-Link Layer

The data-link layer is responsible for transferring data between the ends of a physical link. It must also detect errors, create blocks made up of bits, called frames, and control data flow to reduce congestion. Additionally, the data-link layer must correct problems resulting from damaged, lost, or duplicate frames. The main function of this layer is switching.

3.1.1.1.3 Network Layer

The network layer provides the means for connecting and delivering data from one end to another. It also controls interoperability problems between intermediate networks. The main function performed at this layer is routing.

3.1.1.1.4 Transport Layer

The transport layer receives data from the upper layers, divides it in smaller units if necessary, transfers it to the network layer, and ensures that all the information arrives correctly at the other end. Connection between two applications located in different machines takes place at this layer, for example, customer–server connections through application logical ports.

3.1.1.1.5 Session Layer

This layer provides services when two users establish connections. Such services include dialog control, token management (prevents two sessions from trying to perform the same operation simultaneously), and synchronization.

3.1.1.1.6 Presentation Layer

The presentation layer takes care of syntaxes and semantics of transmitted data. It encodes and compresses messages for electronic transmission. For example, one can differentiate a device that works with ASCII coding and one that works with binary-coded decimal (BCD), even though in both cases the information transmitted is identical.

3.1.1.1.7 Application Layer

Protocols of applications commonly used in computer networks are defined in the application layer. Applications found in this layer include Internet surfing applications (Hypertext Transfer Protocol (HTTP)), file transfer (File Transfer Protocol (FTP)), voice-over networks (Voice-over-IP (VoIP)), videoconferences, etc.

3.1.1.2 TCP/IP Reference Model

The set of TCP/IP protocols allows communication between different machines that execute completely different operating systems. This model was developed mainly to solve interoperatibility problems between heterogeneous networks, allowing that hosts need not know the characteristics of intermediate networks.

The following is a description of the four layers of the TCP/IP model.

3.1.1.2.1 Network Interface Layer

The network interface layer connects the equipment to the local network hardware, connects with the physical medium, and uses a specific protocol to access the medium. This layer performs all the functions of the first two layers in the OSI model.

3.1.1.2.2 Internet Layer

This is a service of datagrams without connection. Based on a metrics, the Internet layer establishes the routes that the data will take. This layer uses IP addresses to locate equipment on the network. It depends on the routers, which resend the packets over a specific interface depending on the IP address of the destination equipment. Is equivalent to layer 3 of the OSI model.

3.1.1.2.3 Transportation Layer

The transportation layer is based on two protocols: the User Datagram Protocol (UDP) and the Transmission Control Protocol (TCP).

The UDP is a protocol that is not connection oriented: it provides nonreliable datagram services (there is no end-to-end detection or correction of errors), it does not retransmit any data that has not been received, and it requires little overcharge. For example, this protocol is used for real-time audio and video transmission, where it is not possible to perform retransmissions due to the strict delay requirements one has in these cases.

The TCP is a connection-oriented protocol that provides reliable data transmission, guarantees exact and orderly data transfer, retransmits data that has not been received, and provides guarantees against data duplication. TCP does the task of layer 4 in the OSI model. All applications working under this protocol require a value called port, which identifies an application in an entity and through which the connection is made with another application in another entity. TCP supports many of the most popular Internet applications, including HTTP, Simple Mail Transfer Protocol (SMTP), FTP, and Secure Shell (SSH).

3.1.1.2.4 Application Layer

The application layer is similar to the OSI application layer, serving as a communication interface, providing specific application services. There are many protocols in this layer, among them FTP, HTTP, Internet Message Access Protocol (IMAP), Internet Relay Chat (IRC), Network File System (NFS), Network News Transport Protocol (NNTP), Network Time Protocol (NTP), Post Office Protocol version 3 (POP3), SMTP, Simple Network Management Protocol (SNMP), SSH, Telnet, etc.

3.1.2 Classification of Computer Networks Based on Size

Computer networks may be classified in several ways according to the context of the study being conducted. The following is a classification by

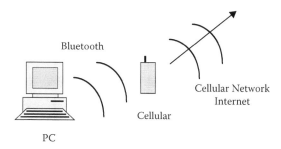

Figure 3.1 PAN design.

size. Later we will specify to which of the networks we will apply the concepts of this book.

3.1.2.1 Personal Area Networks (PANs)

Personal area networks are small home computer networks. They generally connect the home computers to share other devices, such as printers, stereo equipment, etc. Technologies such as Bluetooth are included in PANs.

A typical example of a PAN is the connection to the Internet through a cellular network. Here, the PC is connected via Bluetooth to the cell phone, and through this cell phone we connect to the Internet, as illustrated in Figure 3.1.

3.1.2.2 Local Area Networks (LANs)

Local area networks generally connect businesses, public institutions, libraries, universities, etc., to share services and resources such as Internet, databases, printers, etc. They include technologies such as Ethernet (in any of its speeds, today reaching up to 10 Gbps), Token Ring, 100VG-AnyLan, etc.

The main structure of a LAN traditionally consists of a switch to which one connects the switches where office PCs are connected. Corporate servers and other main equipment are also connected to the main switch. Traditionally, this main switch can be a third-layer switch or can go through a router connected to the main switch, which connects the LAN to the Internet. This connection from the LAN to the carrier, or Internet service provider (ISP), is called the last mile.

Figure 3.2 shows a traditional LAN design.

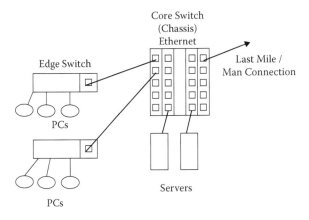

Figure 3.2 LAN design.

3.1.2.3 Metropolitan Area Networks (MANs)

Metropolitan area networks cover the geographical area of a city, interconnecting, for instance, different offices of an organization that are within the perimeter of the same city. Within these networks one finds technologies such as Asynchronous Transfer Mode (ATM), Frame Relay, xDSL (any type of Digital Subscriber Line), cable modem, Integrated Services Digital Network (ISDN), and even Ethernet.

A MAN can be used to connect different LANs, whether among them or with a wide area network (WAN) such as Internet. LANs connect to MANs with what is called the last mile, through technologies such as ATM/Synchronous Digital Hierarchy (SDH), ATM/SONET, Frame Relay/xDSL, ATM/T1, ATM/E1, Frame Relay/T1, Frame Relay/E1, ATM/Asymmetrical Digital Subscriber Line (ADSL), Ethernet, etc. Traditionally, the metropolitan core is made up of high-velocity switches, such as ATM switchboards over an SDH ring or SONET. The new technological platforms establish that MAN or WAN rings can work over Dense Wave Length Division Multiplexing (DWDM); they can go from a 10-Gbps current to transmission velocities of 1.3 Tbps. These high-velocity switches can also be layer 3 equipment, and therefore may perform routing.

Figure 3.3 shows a traditional MAN design.

3.1.2.4 Wide Area Networks (WANs)

Wide area networks span a wide geographical area. They typically connect several local area networks, providing connectivity to devices located in different cities or countries. Technologies applied to these networks are the same as those applied to MANs, but in this case, a larger geographical

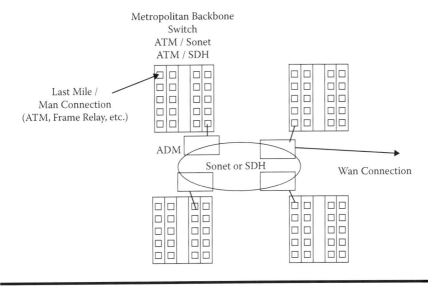

Figure 3.3   MAN design.

area is spanned, and therefore, a larger number of devices and greater complexity in the analysis is required to develop the optimization process.

The most familiar case of WANs is the Internet, as it connects many networks worldwide. A WAN design may consist of a combination of two-layer (switches) and three-layer (routers) equipment, and the analysis depends exclusively on the layer under consideration. Traditionally, in this type of network, it is normally analyzed under a layer 3 perspective.

Figure 3.4 shows a traditional WAN design.

In this book we will work with several kinds of networks. Chapters 4 and 5 discuss MANs and WANs. Said chapters do not include LANs because problems found therein are simpler than problems found in MANs and

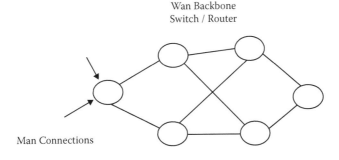

Figure 3.4   WAN design.

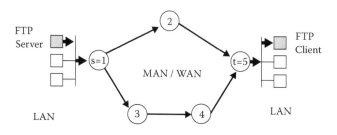

Figure 3.5   Unicast transmission.

WANs. Because Ethernet applies to both LANs and MANs, the same analysis for the WANs can be used for the LANs. Chapter 6 covers wireless networks. Wireless networks do not use physical cables. This type of analysis can be applied on PANs, LANs, MANs, and WANs.

3.1.3 Classification of Computer Networks Based on Type of Transmission

Based on the type of transmission, computer networks fall into the following classifications.

3.1.3.1 Unicast Transmissions

A transmission between a single transmitter and a single receiver is a unicast transmission (Figure 3.5). Examples include applications such as HTTP, FTP, Telnet, VoIP, point-to-point videoconferences, etc.

3.1.3.2 Multicast Transmissions

Transmission of information between one or several transmitters and several receivers is called multicast transmission (Figure 3.6). Applications such as multipoint videoconferences, video streaming, etc., are examples of multicast transmissions.

3.1.3.3 Broadcast Transmissions

Transmission of information between a transmitter and all receivers in a computer network is a broadcast transmission. Usually, these are control messages or messages sent by certain applications for proper operation.

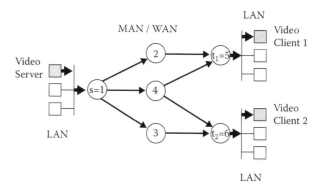

Figure 3.6   Multicast transmission.

This book will focus on unicast and multicast transmissions because they are traditionally the ones intentionally generated by computer network users.

3.2 Computer Network Modeling

A computer network can be represented in a graph where the nodes represent the devices for network interconnection, such as routers, switches, modems, satellites, and others, where the edges represent the physical connection links between such devices. In this book, the nodes will represent the routers or switches (in Internet analysis they would be routers).

The links will represent the media and physical transmission technologies between such routers or switches, for example, optical fiber, copper, satellite links, microwaves, etc. In the case of communications, the nodes could be the satellites and the links, the uplink and downlink.

In the next section we will introduce graph theory to represent and model computer networks as graphs and, consequently, be able to apply different graph algorithms to computer networks.

3.2.1 Introduction to Graph Theory

Graphs are a type of data structure that consists basically of two components: vertices or nodes, and the edges or links that connect them.

Graphs are divided into undirected graphs and directed graphs. Undirected graphs are made up of a set of nodes and a set of links that are

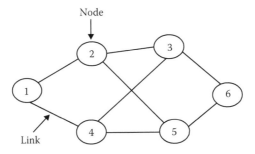

Figure 3.7   Graph.

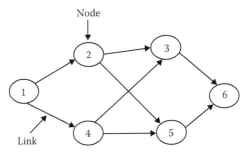

Figure 3.8   Directed graph.

not ordered node pairs. Figure 3.7 shows a undirected graph with six nodes and eight links, in which we refer indistinctly, for example, to link (1, 2) or (2, 1).

Directed graphs are a set of nodes and a set of edges whose elements are ordered pairs of different nodes. Figure 3.8 shows a directed graph (also with six nodes and eight links) with the following links: {(1, 2), (1, 4), (2, 3), (2, 5), (3, 6), (4, 3), (4, 6), (5, 6)}. In this case, we see that there is a link (1, 2) but not a link (2, 1). For link (2, 1) to exist, it would have to be explicit in the set of links.

Because we can represent a computer network through graphs, we can use different algorithms applied to graphs (shortest path, shortest path tree, minimum spanning tree, maximum flow, breadth first search, depth first search, etc.) to optimize resources in the computer networks.

3.2.2 Computer Network Modeling in Unicast Transmission

In the unicast case the network is modeled as a directed graph $G = (N, E)$, where N is the set of nodes and E is the set of links. n denotes the number of network nodes, i.e., $n = |N|$. Among the nodes, we have a source $s \in N$ (ingress node) and a destination $t \in N$ (egress node). Let $(i, j) \in E$ be the link from node i to node j. Let $f \in F$ be any unicast flow, where F is the flow set. We denote the number of flows by $|F|$. Let c_{ij} be the weight of each link (i, j). Let bw_f be the traffic demand of a flow f from the ingress node s to the egress node t.

Figure 3.9a illustrates a computer network for unicast transmissions. We can see that the set of nodes N is {1, 2, 3, 4, 5}, and therefore, $n = |N| = 5$. The set of links E is {(1, 2), (1, 3), (2, 5), (3, 4), (4, 5)}. Because this graph is a directed graph, there is no possibility of transmitting from node 2 to node 1, because link (2, 1) is not a part of set E. $G = (N, E)$ is the graph that represents the network. The nodes, for example, would represent the routers and the links, the fiber optics or the terrestrial or satellite microwave connections. In this example, the origin or transmitter node is node 1 (labeled s) and the destination node or receiver is node 5 s (labeled t).

In this type of network we suppose that origin as well as destination nodes are connected to a LAN such as Ethernet, for example, to which are connected the clients or servers that are transmitting the information. In Figure 3.9b, both the origin node s and the destination node t are connected to an Ethernet LAN, whether through a hub or a switch.

Each link in the networks is assigned a weight. In this example, c_{ij} stands for the weight of each link. The weight can mean a delay in transmission on such link, the linking capacity, the available capacity, the cost of transmitting the information through that link, etc.

Finally, we find that computer networks transmit the traffic produced by certain applications. In this book, when we talk about applications, we generally refer to applications over IP, because this is the transmission protocol on the Internet. For example, these unicast applications can be FTP for file transfers, HTTP for surfing, Telnet to emulate terminals, HTTP-secure, HTTP, SSH-secure Telnet, VoIP for telephone call transfer over IP, videoconferences over IP, video streaming, etc. In this example, we have established a single flow $|F| = 1$, which is an FTP transmission from the

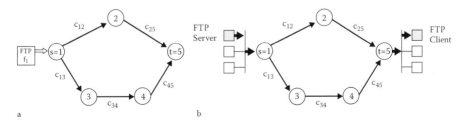

Figure 3.9   Computer network modeling in unicast transmission.

origin node s to destination node t. The FTP flow is shown as f_1, and the set of flows would consist of a single element, that is, $F = \{f_1\}$.

The path through which the flow will be transmitted from the emitting node to the receiving node will be determined by a routing algorithm.

In this section, we have shown how the topology of computer networks transmitting unicast applications through graphs can be represented. In later chapters, this modeling will help us develop the optimization models at the routing level.

3.2.3 Computer Networks Modeling in Multicast Transmission

As we had previously mentioned, there are other types of applications that transmit their data to several destination modes. These transmissions are called multicast transmissions.

In the multicast case, the network can be modeled as a directed graph $G = (N, E)$, where N is the set of nodes and E is the set of links. n denotes the number of network nodes, i.e., $n = |N|$. Among the nodes, we have a source $s \in N$ (ingress node) and some destinations T (the set of egress nodes). Let $t \in T$ be any egress node. Let $(i, j) \in E$ be the link from node i to node j. Let $f \in F$ be any multicast flow, where F is the flow set and T_f is the egress node subset for the multicast flow f. We use $|F|$ to denote the number of flows. Note that $T = \bigcup_{f \in F} T_f$.

Figure 3.10a shows a computer network for multicast transmission. In this figure we can see that the set of nodes N is $\{1, 2, 3, 4, 5, 6\}$, and therefore, $n = |N| = 6$. The set of links e is $\{(1, 2), (1, 3), (1, 4), (2, 5), (3, 6), (4, 5), (4, 6)\}$. The graph that represents the networks is $G = (N, E)$. In this example, the origin or transmitter is node 1 (labeled s) and the set of destination or receiving nodes consists of nodes 5 and 6 (labeled t_1 and t_2, respectively).

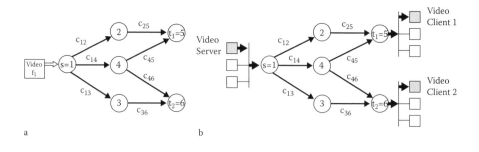

a   b

Figure 3.10 Computer network modeling in multicast transmission.

As in the unicast case, in this type of network we assume that the origin and destination nodes are connected to a LAN such as, for example, Ethernet, to which are connected the clients or servers performing the transmission of information (Figure 3.10b). The origin router s and the destination routers t_1 and t_2 are connected to a LAN Ethernet network through a hub or switch. Like in the unicast case, the weight of each link is represented by c_{ij}.

Examples of multicast applications include multipoint videoconferences, multipoint video streaming, multipoint audio conferences, etc. In this example, we have one single flow $|F| = 1$ that is a video-streaming transmission from origin node s to destination nodes t_1 and t_2. This video-streaming flow we represent as f_1; the set of flows F would consist of a single element $F = \{f_1\}$ and $T = T_f$.

One can see that the modeling of a network for multicast transmission is similar to the unicast case, with the difference that the flow of information will be received by multiple destinations.

Chapter 4

Routing Optimization in Computer Networks

4.1 Concepts

4.1.1 Unicast Case

To solve an optimization problem in computer networks, we need to define a variable that will tell us the route through which the flow of information will be transmitted. We will define the variables vector X_{ij}^f as follows:

$$X_{ij}^f = \begin{cases} 1, \text{ if link } (i, j) \text{ is used for flow } f \\ 0, \text{ if link } (i, j) \text{ is not used for flow } f \end{cases}$$

This means that the variables vector X_{ij}^f is 1 if link (i, j) is used to transmit flow f; if it cannot be used, the vector is 0. In Figure 4.1 we see that there are two possible routes to go from origin node 1 to destination node 5. The first route is through links $(1, 2)$ and $(2, 5)$. The second path is through links $(1, 3)$, $(3, 4)$, and $(4, 5)$. Assuming the first path $((1, 2), (2, 5))$ has been selected as the route through which to transmit flow f_1, $X_{12}^1 = 1$ and $X_{25}^1 = 1$. Because links $(1, 3)$, $(3, 4)$, and $(4, 5)$ are not used, the values of vector X_{ij}^f for these equal 0 (Figure 4.1).

Table 4.1 summarizes the parameters used for the unicast case.

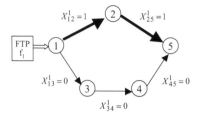

Figure 4.1   Computer network optimization in unicast transmission.

Table 4.1   Parameters Used for the Unicast Case

Terms	Definition
G (N, E)	Graphs of the topology
N	Set of nodes
E	Set of links
s	Ingress node
t	Egress node
(i, j)	Link from node *i* to node *j*
F	The flow set
f	Any unicast flow
X_{ij}^{f}	Indicates whether the link (i, j) is used for flow *f* with destination to the egress node *t*
c_{ij}	The available capacity of each link (i, j)
bw_f	The traffic demand of a flow *f*

4.1.2 Multicast Case

For the multicast case, we must also define a variable that shows us whether the link is used to transmit a specific flow.

We will define the variables vector X_{ij}^{tf} as follows:

$$X_{ij}^{tf} = \begin{cases} 1, \textit{if link } (i, j) \textit{ is used for flow } f \textit{ with destination } t \\ \\ 0, \textit{if link } (i, j) \textit{ is not used for flow } f \textit{ with destination } t \end{cases}$$

This means that variable X_{ij}^{tf} is 1 if link (i, j) is used to transmit multicast flow *f* with destination node *t*, which belongs to the set of destination nodes *Tf*. If it is not used, it is 0. In Figure 4.2 we can see that there are four possible trees to go from origin node 1 to destination nodes 5 and 6. The first tree consists of paths {(1, 2), (2, 5)} and {(1, 3), (3, 6)}. The second tree consists of {(1, 4), (4, 5)} and {(1, 4), (4, 6)}. The third tree

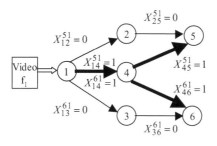

Source: s = 1   Destinations: $t_1 = 5$ and $t_2 = 6$

Figure 4.2   Computer network optimization in multicast transmission.

consists of paths {(1, 2), (2, 5)} and {(1, 4), (4, 6)}. Finally, the fourth tree consists of paths {(1, 4), (4, 5)} and {(1, 3), (3, 6)}. Assuming that the second tree was selected ({(1, 4), (4, 5)} and {(1, 4), (4, 6)}) as the tree to transmit multicast flow f_1, $X_{14}^{51} = 1$ and $X_{14}^{61} = 1$, because both links are used to transmit flow f_1 with destination nodes 5 and 6. Because links (1, 2), (2, 5), (1, 3), and (3, 6) are not being used, the values of vector X_{ij}^{tf} for these links for both destination nodes 5 and 6 (Figure 4.2) are zero. Variables corresponding to the links through which one cannot reach a specific destination are not shown in Figure 4.2; for example, variable $X_{45}^{61} = 0$, because it is impossible to reach node 6 through link (4, 5).

Table 4.2 summarizes the parameters used for the multicast case.

Table 4.2   Parameters Used for the Multicast Case

Terms	Definition
$G(N, E)$	Graphs of the topology
N	Set of nodes
E	Set of links
S	Ingress node
T	Set of egress nodes or any egress node
(i, j)	Link from node i to node j
F	The flow set
f	Any multicast flow
T_f	The egress node subset for the multicast flow f
X_{ij}^{tf}	Indicates whether the link (i, j) is used for flow f with destination to the egress node t
c_{ij}	The available capacity of each link (i, j)
bw_f	The traffic demand of a flow f

4.2 Optimization Functions

In this section we will define certain functions that will later be used for multi-objective optimization. This book will show some of the functions most commonly used, but we are not attempting an exhaustive analysis of all possible functions that can be considered.

4.2.1 Hop Count

4.2.1.1 Unicast Transmission

The first function to be analyzed is hop count. This function represents the number of links through which packets must pass from the time they leave the origin node until they reach the destination node.

Figure 4.3 shows that there are two possible paths to carry the flow of information f_1 (File Transfer Protocol (FTP)) between origin node 1 and destination node 5: The first path is formed by links (1, 2) and (2, 5), and therefore, the number of hops to be taken by the packets is 2. The second path is formed by links (1, 3), (3, 4), and (4, 5) and, in this case, the number of hops is 3.

If we use the number of hops as an optimization function, then the path chosen to transmit the packets from origin node 1 to destination node 5 will be the first path.

The routing protocol that works with the hops function to calculate the shortest path is called the Routing Internet Protocol (RIP). This protocol is used mainly in small metropolitan area networks (MANs) or wide area networks (WANs) or in routing in local area networks (LANs) through the virtual LANs (VLANs).

To mathematically describe the hop count function we will use an example. In Figure 4.3 we have the variables X_{ij}^f describing the paths. If we want to describe the paths mentioned above under these variables, we could say that the number of hops for the first path would be given

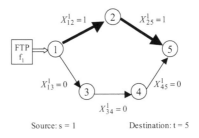

Figure 4.3 Hop count function in unicast transmission.

by $X_{12}^1 + X_{25}^1$. If this is the path used, the value of these variables would be 1 and the sum would yield 2. Similarly, for the second path, the value of the function would be given by $X_{13}^1 + X_{34}^1 + X_{45}^1$, and if this is the path used, the value of each of these variables would be 1, and therefore, the sum would yield 3.

When we want to find the shortest path based on the hop count function, only one path will be selected. For the example shown previously, it would be the path given by links (1, 2) and (2, 5).

The function to be optimized would consist of minimizing the sum of all the paths, but only the variables X_{ij}^f of one path would have a value of 1. Therefore, the hop count function can be stated the following way:

$$\min \sum_{f \in F} \sum_{(i,j) \in E} X_{ij}^f$$

$\sum_{(i,j) \in E} X_{ij}^f$ denotes the possible paths for flow f (of which only one will

be taken). $\sum_{f \in F}$ denotes all the flows that will be transmitted over the

network. In the solution, for every flow f, one will get a path with the minimum number of hops from its origin node s through its destination node t.

4.2.1.2 Multicast Transmission

In this section we will analyze how we can mathematically express the hop count function when we perform multicast flow transmissions, that is, when the information is sent to a set of destination nodes.

Figure 4.4 shows that there are four possible trees to carry the flow of information f_1 (video) between node 1 and destination nodes 5 and 6: The first tree (Figure 4.5) is formed by paths {(1, 2), (2, 5)} and {(1, 3), (3, 6)}, and therefore, the number of total hops for this tree will be the sum of the number of hops in each path. Because there are 2 hops in the first path and also 2 hops in the second path, the total number of hops in the tree is 4. The second tree (Figure 4.6) is formed by paths {(1, 4), (4, 5)} and {(1, 4), (4, 6)}, and therefore, the total number of hops in this tree is 4. The third tree (Figure 4.7) is formed by paths {(1, 2), (2, 5)} and {(1, 4), (4, 6)}, and therefore, the total number of hops is 4. The fourth tree (Figure 4.8) is formed by paths {(1, 4), (4, 5)} and {(1, 3), (3, 6)}, and therefore, the total number of hops is 4.

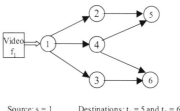

Source: s = 1   Destinations: t_1 = 5 and t_2 = 6

Figure 4.4   Hop count function in multicast transmission.

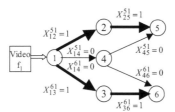

Source: s = 1   Destinations: t_1 = 5 and t_2 = 6

Figure 4.5   First solution.

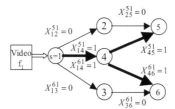

Source: s = 1   Destinations: t_1 = 5 and t_2 = 6

Figure 4.6   Second solution.

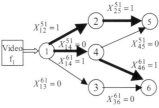

Source: s = 1   Destinations: t_1 = 5 and t_2 = 6

Figure 4.7   Third solution.

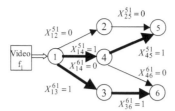

Source: s = 1   Destinations: t_1 = 5 and t_2 = 6

Figure 4.8   Fourth solution.

Because we are using the total number of hops as the optimization function, it is possible to take any of the four trees as the optimal tree. In multicast transmissions, the routing protocol that works based on this hop count function is called the Distance Vector Multicast Routing Protocol (DVMRP).

To mathematically describe the hop count function for the multicast case we will use an example. In the previous figures one can see the variables X_{ij}^{tf} that describe the trees. The number of hops for the first tree would be given by: $\left(X_{12}^{51} + X_{25}^{51} \right) + \left(X_{13}^{61} + X_{36}^{61} \right)$. If this is the tree used, the value of these variables would be 1, and therefore, the sum would be 4. We can calculate for the other three trees in a similar way.

When we want to find the shortest path tree based on the hop count function, only one tree will be selected for transmission of the multicast flow, but any of these four trees could be used.

The function to be optimized would consist of minimizing the sum of all the paths with destination node t. The hop count function for multicast transmission can be stated in the following way:

$$\min \sum_{f \in F} \sum_{t \in T_f} \sum_{(i,j) \in E} X_{ij}^{tf}$$

$\sum_{(i,j) \in E} X_{ij}^{tf}$ denotes one of the possible paths for flow f with destination node t. $\sum_{t \in T_f}$ denotes that one must find a path for each of the nodes t of flow f. $\sum_{f \in F}$ denotes all multicast flows transmitted over the network.

For each flow f one will get a tree with the minimum number of hops from its origin node s through the set of destination nodes T.

4.2.2 Delay

4.2.2.1 Unicast Transmission

There are different kinds of delay that are complementary among each other. Some authors showed that delay has three basic components: switching delay, queuing delay, and propagation delay. The switching delay is a consistent value and can be added to the propagation value. The queuing delay is already reflected in the bandwidth consumption. The authors state that the queuing delay is used as an indirect measure of buffer overflow probability (to be minimized). Other computational studies have shown that it makes little difference whether the cost function used in routing includes the queuing delay or the much simpler form of link utilization.

Because later in this book we will analyze the bandwidth consumption function, we will only analyze the propagation delay here. The unit associated to the delay function is time, and it is usually expressed in milliseconds (ms). Because the objective of this function is to calculate and obtain the minimum delay end to end, this function is cumulative.

Using the same network of previous examples, we will analyze how the delay function can change the result of the shorter path. Figure 4.9 shows a graph for the analysis by means of the delay function. A weight d_{ij} has been added to each link; it denotes the propagation delay between node i and node j. As in the previous example, Figure 4.9 shows that there are two possible paths to carry the flow of information f_1 (FTP)

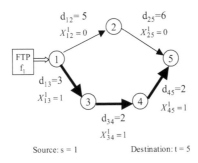

Figure 4.9   Delay function.

between origin node 1 and the destination node 5: The first path is formed by links (1, 2) with a delay d_{12} of 5 ms and (2, 5) with a delay d_{25} of 6 ms. Because this is a cumulative function, the end-to-end delay experienced by the packets during transmission using this path would be 11 ms. The second path is formed by links (1, 3) with a delay of 3 ms, (3, 4) with a delay of 2 ms, and (4, 5) with a delay of 2 ms; the end-to-end delay is 7 ms. If we use the end-to-end delay as the optimization function, the second path will be selected to transmit the packets from origin node 1 to destination node 5.

If we compare this result with the analysis using the hop count function (Figure 4.3), we see that the best path is different in every case. By means of multi-objective optimization, we can find all of these solutions when the functions are in conflict.

Among the unicast routing protocols that can work with this type of function are Open Shortest Path First (OSPF), Interior Gateway Routing Protocol (IGRP), Enhanced Interior Gateway Routing Protocol (EIGRP), Intermediate System to Intermediate System (IS-IS), and Border Gateway Protocol (BGP).

To mathematically describe the end-to-end delay function we will use an example (Figure 4.9). In this case, the delay for the first path would be given by $d_{12}.X_{12}^1 + d_{25}.X_{25}^1$. If this is the path used, the variables $X_{12}^1 = 1$ and $X_{25}^1 = 1$ would be 1, and therefore, the end-to-end delay for this path would be $(5*1) + (6*1) = 11$ ms. Likewise, for the second path the value of the function would be given by $d_{13}.X_{13}^1 + d_{34}.X_{34}^1 + d_{45}.X_{45}^1$. If this is the path used, the value of these variables X_{ij}^f would be 1; therefore, the calculation would be $(3*1) + (2*1) + (2*1) = 7$ ms.

When we want to find the shortest path based on the end-to-end delay function, only one path will be selected (the second path in this example.) Figure 4.9 shows the solution found, and therefore, the values of variables X_{ij}^f of the second path are 1, while for the other path the value of these variables is 0.

The function to be optimized would consist of minimizing the sum of the product of the delay d_{ij} of each link (i, j) times the value associated with variable X_{ij}^f. The function to minimize the end-to-end delay can be stated the following way:

$$\min \sum_{f \in F} \sum_{(i,j) \in E} d_{ij}.X_{ij}^f$$

$\sum_{(i,j)\in E} d_{ij}.X_{ij}^f$ denotes the end-to-end delay value of the possible paths for

flow f (only one will be taken). $\sum_{f\in F}$ denotes all flows that will be

transmitted over this network. For each flow f, one will get a path with the minimum end-to-end delay from its origin node s to its destination node t.

4.2.2.2 Multicast Transmission

In this section we will analyze how we can mathematically express the end-to-end delay function when we transmit multicast flows.

Figure 4.10 shows that there are four possible trees to carry the flow of information f_1 (video) between node 1 and destination nodes 5 and 6.

Figure 4.11 to Figure 4.14 show the same four possible trees shown previously in the hop count function to carry flow of information f_1 (video) between origin node 1 and destination nodes 5 and 6. In this case, for each tree the total end-to-end delay would be 8 ms (Figure 4.11), 20 ms (Figure 4.12), 14 ms (Figure 4.13), and 14 ms (Figure 4.14), respectively. Therefore, in this case, the solution would be given by the tree formed by paths {(1, 2), (2, 5)} and {(1, 3), (3, 6)}.

Other multicast routing protocols that can work with this type of function are, among others, Protocol Independent Multicast — Dense Mode (PIM-DM), Multicast Open Shortest Path First (MOSPF), and Border Gateway Multicast Protocol (BGMP).

In this book we do not analyze the analytical model for the case of transmission with shared trees, which is how the Protocol Independent Multicast — Sparse Mode (PIM-SM) works, but with some minor changes, it could be implemented. In this case, the change would consist of creating a unicast shortest path (would be the same model of the delay function

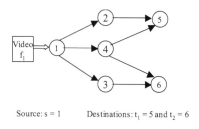

Source: s = 1   Destinations: t₁ = 5 and t₂ = 6

Figure 4.10 Delay function in multicast transmission.

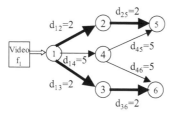

Source: s = 1   Destinations: $t_1 = 5$ and $t_2 = 6$

Figure 4.11 First solution.

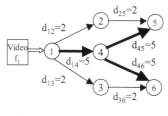

Source: s = 1   Destinations: $t_1 = 5$ and $t_2 = 6$

Figure 4.12 Second solution.

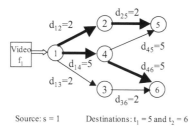

Source: s = 1   Destinations: $t_1 = 5$ and $t_2 = 6$

Figure 4.13 Third solution.

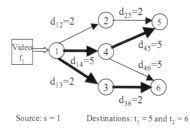

Source: s = 1   Destinations: $t_1 = 5$ and $t_2 = 6$

Figure 4.14 Fourth solution.

in the unicast case) between origin node *s* and node RP (rendezvous point), which works as the root of the shared tree and, subsequently, performs the shortest path tree (would be the same model of the delay function for the multicast case) between node RP and every one of the destination nodes T_f.

To mathematically describe the end-to-end delay function we will use an example (Figure 4.11). In this case, the total delay would be given by $\left(d_{12}.X_{12}^{51} + d_{25}.X_{25}^{51}\right) + \left(d_{13}.X_{13}^{61} + d_{36}.X_{36}^{61}\right)$. If this tree is used, the value of the variables would be 1, and therefore, the value would be ((2*1) + (2*1)) + ((2*1) + (2*1)) = 8 ms. The calculation for the other three trees can be done in the same way.

The function to be optimized would consist of minimizing the sum of all the paths. The function to minimize the end-to-end delay for multicast transmission can be stated as follows:

$$\min \sum_{f \in F} \sum_{t \in T_f} \sum_{(i,j) \in E} d_{ij}.X_{ij}^{tf}$$

4.2.3 Cost

4.2.3.1 Unicast Transmission

The cost of the transmission function may be correlated to the delay function. In this section, we will analyze the cost function because there will be cases where the cost of transmission will not necessarily be correlated with other objective functions such as delay or hops. The analysis of the cost function is similar to the analysis performed with the delay function.

The unit of the cost function is associated with money; for example, the cost of each link may be associated with the cost of using the links in a clear channel transmission with Point-to-Point Protocol (PPP) or High-Level Data-Link Control (HDLC) technology, or may be associated with the use of a commuted operator through a virtual circuit with Asynchronous Transfer Mode (ATM) technology or ATM or Frame Relay, or through a local service provider (LSP) in Multi-Protocol Label Switching (MPLS) technology. Because the objective of this function is to calculate and obtain the minimum end-to-end cost, this function is also cumulative.

Figure 4.15 shows a network to perform the analysis by means of the cost function. In this figure one sees that a weight w_{ij}, which denotes the cost of transmitting any flow between nodes *i* and *j*, has been added to

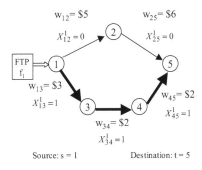

Figure 4.15 Cost function.

every link. As in the case of the delay, in Figure 4.15 one can see that there are two possible paths to carry the flow of information f_1 (FTP) between origin node 1 and destination node 5. The cost of the first path would be $11, and the cost of the second would be $7. If we use the cost of using the links as the optimization function, the second path would be selected to transmit the packets from the origin node 1 through destination node 5.

If we want to describe the above-mentioned paths mathematically, we can say that the cost for the first path would be given by $w_{12}.X_{12}^1 + w_{25}.X_{25}^1$. If this is the path used, the value of variables X_{12}^1 and X_{25}^1 would be 1, and therefore, the total cost for this path would be (5*1) + (6*1) = $11. Likewise, for the second path the value of the function would be given by $w_{13}.X_{13}^1 + w_{34}.X_{34}^1 + w_{45}.X_{45}^1$. If this is the path used, the value of variables X_{13}^1 and X_{34}^1 would be 1, and therefore, the total cost for this path would be (3*1) + (2*1) + (2*1) = $7.

When we want to find the shortest path based on the cost function, only one path will be selected (the second path in the example). Figure 4.15 shows the solution found when cost is minimized, and therefore, the values of the variables X_{ij}^f of this path are 1, while for the other path the value of these variables is 0.

The function to be optimized would consist of minimizing the sum of the product of the cost w_{ij} of every link (i, j) times the associated value of variable X_{ij}^f. The function to minimize cost may be stated in the following way:

$$\min \sum_{f \in F} \sum_{(i,j) \in E} w_{ij}.X_{ij}^f$$

$\displaystyle\sum_{(i,j)\in E} w_{ij}.X_{ij}^{f}$ denotes the end-to-end cost of the possible paths for flow f

(only one will be taken). $\displaystyle\sum_{f\in F}$ denotes all flows that will be transmitted over

this network. For every flow f, one will obtain a path with the minimum end-to-end cost from its origin node s through its destination node t.

4.2.3.2 Multicast Transmission

The cost function will be analyzed when we present the bandwidth consumption function because these functions exhibit similar behaviors when multicast transmissions are performed.

4.2.4 Bandwidth Consumption

4.2.4.1 Unicast Transmission

The bandwidth consumption function represents the bandwidth that is used in every link throughout the path to transmit the flow of information from origin node s through destination node t. One needs to mention here that the function analyzed in this item refers to the bandwidth consumed in the whole network to carry the flow of information, and not the analysis of how much bandwidth is available in a path, whose function would be denoted by another mathematical expression.

The bandwidth consumption function is associated with the transmission capacity used in every one of the links, and its values are given in bits per second (bps). It is common that the capacities of the links in low-speed WAN or MAN channels are in kilobits per seconds (Kbps). In high-speed WAN and MAN links and in LANs it is common to use magnitudes in the order of megabits per seconds (Mbps) or gigabits per second (Gbps). Finally, in the new Generalized Multiprotocol Label Switching (GMPLS) networks using Dense Wavelength Division Multiplexing (DWDM), it is possible to obtain measurements of the order of terabits per second (Tbps).

Because the objective of this function consists of calculating and obtaining the minimum bandwidth consumption in the whole path, this function is cumulative. If we analyze the available bandwidth function, this function is concave.

Figure 4.16 shows a network to perform the analysis through the bandwidth consumption function. Contrary to the networks used for the previous function, link (1, 5) has been added in this figure. Variable c_{ij} denotes the available capacity in the link (i, j). One can see in Figure

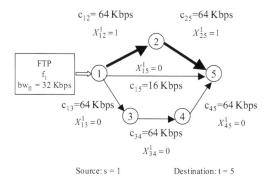

Figure 4.16 Bandwidth consumption function.

4.16 that there are three possible paths to carry the flow of information f_1 (FTP) between origin node 1 and destination node 5. In this example, the first path is formed by links (1, 2), with a capacity of 64 Kbps, and (2, 5), with a capacity of 64 Kbps. Because this function is cumulative, the application has a traffic demand bw_{f1} of 32 Kbps, and because it must pass through the two links, the bandwidth consumption when transmitting the flow through this path would be 64 Kbps: 32 Kbps consumed by (1, 2) plus 32 Kbps consumed by (2, 5). The second path consists only of link (1, 5), but there is a drawback to this link: it cannot transmit flow f_1 because flow f_1 (bw_{f1}) demands a capacity of 32 Kbps and the capacity of channel c_{15} is only 16 Kbps. The third path is formed by links (1, 3), with a capacity of 64 Kbps, (3, 4), with a capacity of 64 Kbps, and (4, 5), with a capacity of 64 Kbps; as for the first path, the value of the bandwidth consumption in the complete path would be 96 Kbps.

If we use bandwidth consumption as the optimization function, the first path would be selected to transmit the packets from origin node 1 to destination node 5, with a bandwidth consumption of 64 Kbps.

Comparing this result obtained with the means of the hop count function (see Figure 4.3), it could be that both functions are correlated. However, in this last example one can see that if we only minimize the hop count function, the result is feasible because the second path, even though it only has 1 hop, does not have enough capacity to carry the demand of flow f_1.

When we analyze this same function in multicast transmission, we will see that it will be very useful due to the nature of the multicast transmission of creating tree transmission.

To mathematically describe the bandwidth consumption function we will use an example (Figure 4.16). In this case, the bandwidth consumption for the first path would be given by $bw_1.X_{12}^1 + bw_1.X_{25}^1$. If this is the path used, the value of the variables X_{12}^1 and X_{25}^1 would be 1, and therefore, the value for this path would be (32*1) + (32*1) = 64 Kbps. Similarly, for the second path the value of the function would be given by $bw_1.X_{15}^1$. If this is the path used, the value of variable X_1^1 would be 1, and therefore, the calculation would be 32*1 = 32 Kbps. But because the capacity of this channel is only 16 Kbps, this path is not a feasible solution for the problem. For the third path, the value of the function would be given by $bw_1.X_{13}^1 + bw_1.X_{34}^1 + bw_1.X_{45}^1$; now, if this is the path used, the value of the variables X_{13}^1, X_{34}^1, and X_{45}^1 would be 1, and therefore, the value would be (32*1) + (32*1) + (32*1) = 96 Kbps.

When we want to find the shortest path based on the bandwidth consumption function, only one path will be selected, and for the example previously shown, it would be the first path given by links (1, 2), and (2, 5). Figure 4.16 shows the solution found when minimizing bandwidth consumption, and therefore, the values of variables X_{ij}^f of this path are 1, while for the other paths the value of these variables is 0.

The function to be optimized would consist of minimizing the sum of the product of the demand bw_f for each flow f times the value associated with variable X_{ij}^f and can be expressed in the following way:

$$\min \sum_{f \in F} \sum_{(i,j) \in E} bw_f.X_{ij}^f$$

$\sum_{(i,j) \in E} bw_f.X_{ij}^f$ denotes the value of the bandwidth consumption of the

possible paths for flow f (only one will be taken). $\sum_{f \in F}$ denotes all flows

that will be transmitted over this network. For each flow f, one will obtain a path with the minimum bandwidth consumption from its origin node s through its destination node t.

4.2.4.2 Multicast Transmission

This function makes an interesting contribution in multicast transmission. As discussed in previous chapters for multicast transmissions, when one wants to send a packet to more than one destination, if packages whose destinations are different pass through the same link, only one is transmitted through such link. The router will copy this package to be able to transmit and resend it through more than one exit port.

Figure 4.17 shows that there are four possible trees to transmit flow f_1 from origin node s to destination nodes t_1 and t_2. The first tree would

$$c_{ij} = 64\,Kbps, \forall(i, j) \in E$$

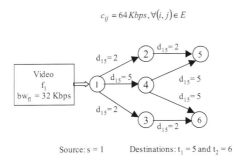

Source: s = 1   Destinations: t_1 = 5 and t_2 = 6

Figure 4.17   Bandwidth consumption function in multicast transmission.

be formed by paths {(1, 2), (2, 5)} and {(1, 3), (3, 6)}. As we have seen previously, functions hops, total delay, and bandwidth consumption would give the following values: 4 hops, 8-ms delay, and 128-Kbps bandwidth consumption, respectively. In this case, calculation of the bandwidth consumption function would be given by 32 Kbps of flow f_1 passing through links (1, 2) and (2, 5) to reach destination t_1, and another 32 Kbps passing through links (1, 3) and (3, 6) to reach destination t_2. Therefore, total consumption would be the sum of the demand of flow f_1 (32 Kbps) passing through the four links, that is, 32 Kbps * 4 = 128 Kbps.

The second tree would be formed by paths {(1, 4), (4, 5)} and {(1, 4), (4, 6)}. In this tree the values would be 4 hops for the hop count and 20 ms for the delay. But because the transmission is multicast, only one packet would pass through link (1, 4) to destinations t_1 and t_2, instead of one for each of the destinations, and therefore, the bandwidth consumption function in link (1, 4) would only be 32 Kbps. These packets would arrive at nodes t_1 and t_2 because at node 4, such packets would be copied to exit through links (4, 5) and (4, 6), consuming 32 Kbps at each of them. Therefore, the value of the bandwidth consumption function when transmitting such flow through this tree would be 32 Kbps * 3 = 96 Kbps. If we compare this tree with the first tree, we can see that for the delay function, the first tree is better; for the hops count function, both trees have the same value; and for the bandwidth consumption function, the second tree shows a better value than the first. At the multi-objective optimization level these two trees would be noncomparable solutions.

The third tree would be formed by paths {(1, 2), (2, 5)} and {(1, 4), (4, 6)}. In this case, the values of the functions would be 4 hops, 14 ms, and 128 Kbps.

Finally, the fourth tree would be formed by paths {(1, 4), (4, 5)} and {(1, 3), (3, 6)}, and the values would be 4 hops, 14 ms, and 128 Kbps.

We can see that if we minimize this function, one tends to find trees with paths that intercept each other to reach the destination nodes, and this way profit from the functionality of the multicast transmissions.

If we optimize a unicast transmission, the function can be expressed in the following way:

$$\min \sum_{f \in F} \sum_{t \in T_f} \sum_{(i,j) \in E} bw_f . X_{ij}^{tf}$$

But in the case of a multicast transmission, the mathematical formula would be expressed differently. Once again, let us take the second tree that we previously analyzed, which showed the best value for the bandwidth consumption function. Variables $\left\{ X_{14}^{51}, X_{14}^{61}, X_{45}^{51}, X_{46}^{61} \right\}$ X_{ij}^{tf} would be 1, and the value of the rest of the variables X_{ij}^{tf} would be 0. In this case, variables X_{14}^{51} and X_{14}^{61} would both have a value of 1, showing that the multicast packets of flow f_1 that travel to destinations 5 and 6 are passing through the same link, (1, 4). But, as we had mentioned previously, duplicate packets do not pass through this link, and we can use the *max* function to determine the maximum value of the variables that pass through this link for the same multicast flow, but for different destinations. Hence, the value would be given by $\max \left\{ X_{14}^{51}, X_{14}^{61} \right\} = \left\{ 1, 1 \right\} = 1$. By means of the previous scheme, we would mathematically express in the model that the packets are not transmitted in duplicate for every link that shares a same multicast flow. Therefore, the bandwidth consumption function for multicast transmissions can be restated in the following way:

$$\min \sum_{f \in F} \sum_{(i,j) \in E} bw_f . \max \left(X_{ij}^{tf} \right)_{t \in T_f}$$

If we analyze the cost function, the behavior would be similar to the bandwidth consumption function because the same link is used to reach several destinations; only one packet is transmitted, and the cost would be associated to the transmittal of such packet. Therefore, the cost function in multicast transmissions can be expressed as

$$\min \sum_{f \in F} \sum_{(i,j) \in E} w_{ij} . \max \left(X_{ij}^{tf} \right)_{t \in T_f}$$

where w_{ij} is the cost of (i, j).

4.2.5 Packet Loss Rate

Packet loss rate in transmission consists of calculating and obtaining the minimum end-to-end packet loss rate. Being a probabilistic value at each link, this function is multiplicative.

Figure 4.18 shows a network to perform the analysis through the packet loss rate function. In this figure one can see that the variable plr_{ij}, which is between nodes i and j, has been added to each link. As in the previous case, in Figure 4.18 one can see that there are two possible paths to carry the flow of information f_1 (FTP) between the origin node 1 and the destination node 5.

The first path is formed by link (1, 2), with a packet loss rate plr_{12} of 0.5, and link (2, 5), with a packet loss rate plr_{25} of 0.6. Because it is a multiplicative function, the packet loss rate that the packets would experience when transmitted through this path would be 0.6*0.5 = 0.3.

The second path is formed by links (1, 3), with a packet loss rate of 0.2; (3, 4), with a packet loss rate of 0.2; and (4, 5), with a packet loss rate of 0.2, and in this case, the end-to-end value of the packet loss rate would be 0.2*0.2*0.2 = 0.008.

If we use the end-to-end packet loss rate as the optimization function, the second path would be selected to transmit the packets from origin node 1 through destination node 5.

To mathematically describe the end-to-end packet loss rate function we will use an example (Figure 4.18). In this example, the packet loss rate for the first path would be given by $plr_{12}.X_{12}^1 * plr_{25}.X_{25}^1$. If this is the path used, the values of X_{12}^1 and X_{25}^1 would be 1, and therefore, the calculation for this path would be (0.5*1)*(0.6*1) = 0.3. Similarly, for the second path, the value of the function would be given by $plr_{13}.X_{13}^1 * plr_{34}.X_{34}^1 * plr_{45}.X_{45}^1$; now, if this is the path used, the value

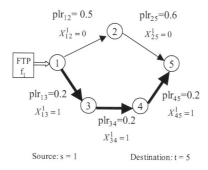

Figure 4.18   Packet loss rate function.

of these X_{ij}^f variables would be 1; hence, the calculation would be $(0.2*1)*(0.2*1)*(0.2*1) = 0.008$.

When we want to find the shortest route based on the packet loss rate function, only one path will be selected, and in the example previously shown, it would be the second path, given by links (1, 3), (3, 4), and (4, 5). Figure 4.18 shows the solution when minimizing packet loss rate, and therefore, the values of variables X_{ij}^f of this path are 1, whereas for the other path the value of such variables is 0.

The function to minimize packet loss rate may be expressed in the following way:

$$\min \sum_{f \in F} \prod_{(i,j) \in E} plr_{ij}.X_{ij}^f$$

$\prod_{(i,j) \in E} plr_{ij}.X_{ij}^f$ denotes the end-to-end packet loss rate value of the possible paths for the flow f (only one will be taken).

The packet loss rate at every link may be calculated in the following way: if $Packet_{tx\_i}$ is the number of packets transmitted through node i and $Packet_{rx\_j}$ is the number of packets received through node j using link (i, j), we can express packet loss rate at link (i, j) as

$$plr_{ij} = \frac{Packet_{rx\_j}}{Packet_{tx\_i}}$$

4.2.6 Blocking Probability

4.2.6.1 Unicast Transmission

The next function that we can optimize is the blocking probability function. This function behaves like the packet loss function with the difference that while in the packet loss function one analyzes the packets received versus packets transmitted, in the blocking probability function one analyzes the number of connections (flows f) that can actually be transmitted versus the total number of flows requesting transmission ($|F|$).

To calculate the value of the blocking probability function, we can establish the following definitions. If $Connection_{total} = |F|$ is the total number of connections that one wishes to establish and $Connection_{real}$ is the amount of connections that can actually be established due to the network's bandwidth resources, we can express the blocking probability as $1 - BP$, where

$$BP = \frac{Connection_{real}}{connection_{total}}$$

The expression $Connection_{real}$ can be defined as $Connection_{real} = \sum_{f \in F} \max\left(X_{ij}^f\right)_{(i,j) \in E}$ as a function of variables X_{ij}^f. Optimization of this function would be associated with the maximum flow theory.

The max function calculates the maximum value for the variables X_{ij}^f in the path from origin node s through destination node t for every flow f. If for a flow f there is a path to transmit such flow, the value of the corresponding variables X_{ij}^f will be 1, and therefore, the value of $\max\left(X_{ij}^f\right)_{(i,j) \in}$ will also be 1. But if for a flow f one cannot build a path due to lack of network broadband width availability, the value of the variables X_{ij}^f will be 0, and therefore, the value of $\max\left(X_{ij}^f\right)_{(i,j) \in}$ will also be 0. $\sum_{f \in F}$ denotes all the flows that will be transmitted over the network, and therefore will tell us how many flows can actually be transmitted with the current characteristics of the bandwidth resources. Because $Connection_{real} \leq Connection_{total}$, $\sum_{f \in F} \max\left(X_{ij}^f\right)_{(i,j) \in E} \leq |F|$, and therefore, the function to optimize may be

$$\min 1 - BP, \text{ where } BP = \frac{Connection_{real}}{connection_{total}}$$

or

$$\max Connection_{real} = \sum_{f \in F} \max\left(X_{ij}^f\right)_{(i,j) \in E}$$

4.2.6.2 Multicast Transmission

For this function we can refer to the analysis performed on unicast transmissions. The function to be optimized would consist of maximizing the number of connections, that is, the number of flows f that can be transmitted on the network, and the function to minimize the probability of blocking for multicast transmissions can be expressed in the following way:

$$\min 1 - BP, \text{ where } BP = \frac{Connection_{real}}{connection_{total}}$$

or

$$\max Connection_{real} = \sum_{f \in F} \max\left(X_{ij}^{tf}\right)_{(i,j)\in, t\in T_f}$$

where $Connection_{total} = |F|$ would be equally defined in the unicast transmission.

4.2.7 Maximum Link Utilization

4.2.7.1 Unicast Transmission

When we analyze the bandwidth consumption function we see that the path given by link {(1, 5)} would be a nonfeasible solution because the capacity of the link was 16 Kbps and the demand of flow f_1 was 32 Kbps. However, paths {(1, 2), (2, 5)} and {(1, 3), (3, 4), (4, 5)} were feasible solutions because the capacity of all their links was 64 Kbps, and the flow f_1 could be transmitted through them.

But if the origin node wishes to transmit an application f_1 at a rate of 128 Kbps (Figure 4.19), there would not be a feasible solution on this network, and the problem would not have a solution.

If we somehow split flow f_1 into two subflows of, for example, 64 Kbps, one subflow could be transmitted through path {(1, 2), (2, 5)} and the other subflow through path {(1, 3), (3, 4), (4, 5)}, and then there would be a feasible solution. It is important to note that for the split considered, path {(1, 5)} was not considered.

This technique is called load balancing, and it is precisely through the maximum link utilization function that one can obtain solutions of this type.

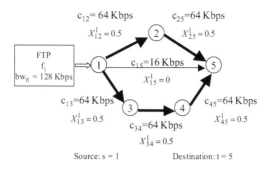

Source: s = 1   Destination: t = 5

Figure 4.19 Maximum link utilization function.

In this case, the vector of variables $X_{ij}^f \in \mathfrak{R}, X_{ij}^f \in [0,1]$ will indicate the fraction of flow f that is transmitted through link (i, j).

To mathematically describe the maximum link utilization function we will use an example (Figure 4.19). Here, the maximum link utilization for the first path would be given by $\max\left\{\left(bw_1.X_{12}^1/c_{12}\right),\left(bw_1.X_{25}^1/c_{25}\right)\right\}$. If this is the path used, the value of variables X_{12}^1 and X_{25}^1 would be 0.5 because paths in this example transmit up to 50 percent of the demand of flow f_1, and therefore, the value for this path would be max{(128*0.5/64), (128*0.5/64)} = 1. Similarly, for the second path, the value of the function would be given by $\max\left\{\left(bw_1.X_{13}^1/c_{13}\right),\left(bw_1.X_{34}^1/c_{34}\right),\left(bw_1.X_{45}^1/c_{45}\right)\right\}$. If this is the path used, the value of variables X_{13}^1, X_{34}^1, and X_{45}^1 would also be 0.5, and therefore, the value would be max{(128*0.5/64), (128*0.5/64), (128*0.5/64)} = 1. Because no others are feasible, because none of the paths have enough capacity to transmit the demand of flow f_1, which is 128 Kbps, the maximum link utilization for the whole network would be

$\alpha = \max\left(MLU\_path1, MLU\_Path2\right) = \max(1,1) = 1$, where MLU is maximum link utilization.

In this solution the links of paths {(1, 2), (2, 5)} and {(1, 3), (3, 4), (4, 5)} are used completely, and therefore, MLU = 1 in them. However, link (1, 5) has not been used, the value of variable X_{15}^1 would be 0, and, consequently, MLU = 0.

The function to be optimized would consist of minimizing   (the maximum utilization in the links of the whole network). This function may be expressed in the following way:

$$\min \quad \alpha$$

where

$$\alpha = \max\left\{\alpha_{ij}\right\}, \text{ where } \alpha_{ij} = \frac{\displaystyle\sum_{f \in F} bw_f . X_{ij}^f}{c_{ij}}$$

$\displaystyle\sum_{f \in F} bw_f . X_{ij}^f$ denotes the value of total traffic demand for all flows $f \in F$ that are transmitted through link (i, j). When one divides the previous value by the capacity of this link (i, j), we obtain the percentage of usage on link (i, j), which is denoted by α_{ij}. Because the objective is to minimize the maximum usage value, we define $\alpha = \max\left\{\alpha_{ij}\right\}$.

Suppose we have a link (i, j) with a capacity C_{ij} (Figure 4.20a). When the maximum link utilization is used, one establishes a new upper limit $(\alpha \cdot c_{ij})$ in this link so that it does not use its full capacity (Figure 4.20b). This way, when the new upper limit is surpassed, the flow of information will be transmitted through other paths instead of using the total capacity of this link. This way we will be performing a load-balancing function.

4.2.7.2 Multicast Transmission

To analyze this function we can resort to the bandwidth consumption function in multicast transmissions and maximum link utilization in unicast transmission. In multicast transmissions, we had expressed the bandwidth consumption function as $\min \displaystyle\sum_{f \in F} \sum_{(i,j) \in E} bw_f . \max\left(X_{ij}^{tf}\right)_{t \in T_f}$. In unicast transmissions, we had expressed the maximum link utilization function as $\min \alpha$, where

 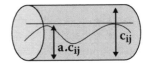

Figure 4.20 **(a) Normal channel. (b) Channel with the function.**

$$\alpha = \max\left\{\alpha_{ij}\right\}, \text{ where } \alpha_{ij} = \frac{\displaystyle\sum_{f \in F} bw_f . X_{ij}^f}{c_{ij}}$$

As had been discussed in the bandwidth consumption function in multicast transmissions, when one uses a same link to send information to more than one destination, only one packet is sent through that link. Therefore, one has to add the function $\max\left(X_{ij}^{tf}\right)_{t \in T_f}$ to the maximum link utilization function for multicast transmissions, through which we can express the previous functioning. Hence, the maximum link utilization function in multicast transmissions would be represented by the following mathematical statement:

$$\min \quad \alpha$$

where

$$\alpha = \max\left\{\alpha_{ij}\right\}, \text{ where } \alpha_{ij} = \frac{\displaystyle\sum_{f \in F} bw_f . \max\left(X_{ij}^f\right)_{t \in T_f}}{c_{ij}}$$

4.2.8 Other Multicast Functions

We can define other functions for multicast transmission that can be based in the functions mentioned above, for example, hop count average, maximal hop count, maximal hop count variation, average delay, maximal delay, maximal delay variation, average cost, maximal cost, etc. These new functions can be associated with the functions analyzed in previous sections, but in real situations they may be very useful for network administrators and designers.

4.2.8.1 Hop Count Average

Hop count average consists of minimizing the average of the number of hops. Because in multicast transmissions there are different paths to reach the destination nodes, the idea would be to minimize the average of the shortest paths to each destination.

This function may be expressed in the following way:

$$\min \frac{Total\_Hop\_Count}{\sum_{f \in F} |T_f|} = \min \frac{\sum_{f \in F} \sum_{t \in T_f} \sum_{(i,j) \in E} X_{ij}^{tf}}{\sum_{f \in F} |T_f|}$$

where $|T_f|$ denotes the standard of T_f, which is the number of destination nodes for flow f.

4.2.8.2 Maximal Hop Count

This function is useful for the maximum requirements of quality of service (QoS).

It may be expressed in the following way:

$$\min \max_{\substack{f \in F, \\ t \in T_f}} \left\{ \sum_{(i,j) \in E} X_{ij}^{tf} \right\}$$

4.2.8.3 Maximal Hop Count Variation

This function is useful when all destination nodes must receive information within a short time interval.

It may be expressed in the following way:

$$\min \max_{f \in F} \left\{ H_f \right\}$$

where

$$H_f = \max_{t \in T_f} \left\{ \sum_{(i,j) \in E} X_{ij}^{tf} \right\} - \min_{t \in T_f} \left\{ \sum_{(i,j) \in E} X_{ij}^{tf} \right\}$$

4.2.8.4 Average Delay

This function consists of minimizing the average of delays to each destination. Because in multicast transmissions there are different paths to reach the destination nodes, the idea is to minimize the average of the paths with the smallest delay.

This function may be expressed in the following way:

$$\min \frac{Total\_Delay}{\sum_{f \in F} |T_f|} = \min \frac{\sum_{f \in F} \sum_{t \in T_f} \sum_{(i,j) \in E} d_{ij}.X_{ij}^{tf}}{\sum_{f \in F} |T_f|}$$

where $|T_f|$ denotes the standard of T_f, that is, the number of destination nodes for flow f.

4.2.8.5 Maximal Delay

This function is useful for the maximum requirements of quality of service (QoS).

It may be expressed in the following way:

$$\min \max_{\substack{f \in F, \\ t \in T_f}} \left\{ \sum_{(i,j) \in E} d_{ij}.X_{ij}^{tf} \right\}$$

4.2.8.6 Maximal Delay Variation

This function is useful when all the destination nodes must receive information within a short time interval.

It may be expressed in the following way:

$$\min \max_{f \in F} \left\{ \Delta_f \right\}$$

where

$$\Delta_f = \max_{t \in T_f} \left\{ \sum_{(i,j) \in E} d_{ij}.X_{ij}^{tf} \right\} - \min_{t \in T_f} \left\{ \sum_{(i,j) \in E} d_{ij}.X_{ij}^{tf} \right\}$$

4.2.8.7 Average Cost

This function consists of minimizing the average of costs to each destination.

It may be expressed in the following way:

$$\min \frac{Total\_Cost}{\sum_{f \in F} |T_f|} = \min \frac{\sum_{f \in F} \sum_{(i,j) \in E} w_{ij}.\max\left(X_{ij}^{tf}\right)_{t \in T_f}}{\sum_{f \in F} |T_f|}$$

4.2.8.8 Maximal Cost

This function is useful for the maximum requirements of quality of service (QoS).

It may be expressed in the following way:

$$\min \max_{f \in F} \left\{ \sum_{(i,j) \in E} \max\left(X_{ij}^{tf}\right)_{t \in T_f} \right\}$$

4.3 Constraints

In this section we will introduce the fundamental constraints that are necessary for the solutions found to be feasible.

4.3.1 Unicast Transmission

The first three constraints are the flow conservation constraints, which guarantee that the solutions obtained are valid paths from the origin s to the destination t. Variable X_{ij}^{f} tells us whether link (i, j) is used by flow f. We will consider this a positive variable when the link leaves the node, and negative in the opposite case.

The first constraint ensures that for every flow f only one path exits from origin node s. The equation that models this constraint is given by the following expression:

$$\sum_{(i,j)\in E} X_{ij}^{f} = 1, f \in F, i = s$$

In Figure 4.21, adding the links that exit origin node $s = 1$, we obtain $X_{12}^{1} + X_{13}^{1} = 1$, and therefore, this constraint is met.

The second constraint ensures that for every flow f only one path reaches the destination node t. The equation that models this constraint is given by the following expression:

$$\sum_{(j,i)\in E} X_{ji}^{f} = -1, i = t, f \in F$$

In Figure 4.21, adding the links that reach destination node $t = 5$, we obtain $-X_{25}^{1} - X_{45}^{1} = -1$, and therefore, this constraint is met.

The third constraint ensures that for every flow f everything that reaches an intermediate node (is not origin node s or destination node t) also exits that node through another link. That is, if for every flow f we add the exit links and subtract the entry links of an intermediate node, the value must be 0. The equation that models this constraint is given by the following expression:

$$\sum_{(i,j)\in E} X_{ij}^{f} - \sum_{(j,i)\in E} X_{ji}^{f} = 0, f \in F, i \neq s, i \neq t$$

In Figure 4.21 the intermediate nodes are nodes 2, 3, and 4. For node 2, if we add its exits and subtract its entries, we obtain $X_{25}^{1} - X_{12}^{1} = 0$.

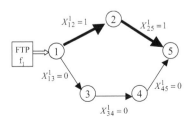

Figure 4.21   Unicast constraints.

For node 3, we would obtain $X_{34}^1 - X_{13}^1 = 0$, and, finally, for node 4, we obtain $X_{45}^1 - X_{34}^1 = 0$. Therefore, this constraint is met for all intermediate nodes.

On the other hand, because in the models shown we have analyzed the demand of traffic necessary for each flow of data and the capacity of the links, one needs to define a constraint to prevent sending a demand of traffic larger than the link's capacity. This constraint can be expressed in the following way:

$$\sum_{f \in F} bw_f . X_{ij}^f \leq c_{ij}, (i, j) \in E$$

$\sum_{f \in F} bw_f . X_{ij}^f$ denotes all the demand for flow (for all flows f) in bits per second (bps) that is transmitted in a link $(i, j) \in E$. The expression $\leq c_{ij}$ does not allow the demand to be higher than the capacity of link c_{ij}.

4.3.2 Multicast Transmission

Just like in unicast transmissions, the first three constraints ensure that the solutions obtained are valid paths from origin s through every one of the destinations $t \, \varepsilon \, T_f$. We will also consider that variable X_{ij}^f is positive when the link exits the node, and negative in the opposite case.

The first constraint ensures that for every destination $t \, \varepsilon \, T_f$ and for flow f, only one path leaves from origin node s. The equation that models this constraint is given by the following expression:

$$\sum_{(i,j) \in E} X_{ij}^{tf} = 1, t \in T_f, f \in F, i = s$$

In Figure 4.22, for example, one has not put variables X_{ij}^{tf} on link (1, 3) with destination node 5, because it is impossible to reach destination node 5 through such a link. In the optimization process, the value of these variables X_{ij}^{tf} would be 0. This constraint is met because adding the exit links for origin node $s = 1$ with destination node $t = 5$, we obtain $X_{12}^{51} + X_{14}^{51} = 1$, and adding the exit links for origin node $s = 1$ with destination node $T = 6$, we obtain $X_{13}^{61} + X_{14}^{61} = 1$.

The second constraint ensures that for every flow f only one path reaches each of the destination nodes t. The equation that models this constraint is given by the following expression:

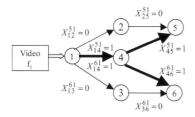

Source: s = 1   Destinations: t_1 = 5 and t_2 = 6

Figure 4.22   Multicast constraints.

$$\sum_{(j,i)\in E} X_{ji}^{tf} = -1, i = t, t \in T_f, f \in F$$

In Figure 4.22, adding the entry links for destination nodes t = 5, we obtain $-X_{25}^{51} - X_{45}^{51} = -1$ and $-X_{36}^{61} - X_{46}^{61} = -1$, and therefore, this constraint is met.

The third constraint ensures that for every flow f, everything that enters an intermediate node also exits such node through another link. This means that if we add the exit links and subtract the entry links in an intermediate node, the value must be 0 for every flow f. The equation that models this constraint is given by the following expression:

$$\sum_{(i,j)\in E} X_{ij}^{tf} - \sum_{(j,i)\in E} X_{ji}^{tf} = 0, t \in T_f, f \in F, i \neq s, i \notin T_f$$

In Figure 4.22 the intermediate nodes are nodes 2, 3, and 4. For node 2, if we add the exits and subtract its entries, we obtain $X_{25}^{51} - X_{12}^{51} = 0$. For node 3, we obtain $X_{36}^{51} - X_{13}^{51} = 0$, and, finally, for node 4, we obtain $X_{45}^{51} - X_{14}^{51} = 0$ and $X_{46}^{61} - X_{14}^{61} = 0$. Therefore, this constraint is met for all intermediate nodes.

With the previous constraints we ensure that for every flow the multicast transmissions generate a single feasible path for its trajectory between the origin node s and every one of the destination nodes t. It is important to note that the set of all these paths creates a tree.

On the other hand, because in the models discussed we have analyzed the traffic demand needed for each flow of data and the capacity of the channels, one needs to define a constraint to avoid sending a traffic demand larger than the capacity of the link. This constraint can be expressed in the following way:

$$\sum_{f \in F} bw_f . \max \left(X_{ij}^{tf} \right)_{t \in T_f} \leq c_{ij}, (i, j) \in E$$

$\sum_{f \in F} bw_f . \max \left(X_{ij}^{tf} \right)_{t \in T_f}$ denotes the complete demand of flow (for all

multicast flows f) in bps that is transmitted in a link $(i, j) \in E$. This mathematical expression was analyzed in the bandwidth consumption function for multicast transmissions. The expression $\leq c_{ij}$ does not allow this flow demand to exceed the capacity of link c_{ij}.

It may happen that on certain jobs some optimization functions will be used as constraints. This is true of the maximum delay function or the maximum hops function. But because the purpose of this book is to perform a multi-objective analysis, these functions will be used exclusively as objective functions and not as constraints.

4.4 Functions and Constraints

In this section we will summarize the functions and constraints discussed above.

4.4.1 Unicast Transmissions

Table 4.3 is a summary of the functions previously defined for the unicast transmissions case.

4.4.2 Multicast Transmissions

Table 4.4 is a summary of the functions previously defined for the multicast transmissions case.

4.5 Single-Objective Optimization Modeling and Solution

In this section we will introduce varied models to solve multi-objective problems using traditional methods that convert the multi-objective problem into a single-objective approximation. In the first cases, we will show how the single-objective approximation is formulated through the different methods, as well as the source code to solve such problem through a solver. For problems with more functions we will only use the weighted

Table 4.3 Objective Functions

	Function	Expression
f_1	Hop count	$$\sum_{f \in F} \sum_{(i,j) \in E} X_{ij}^f$$
f_2	End-to-end delay	$$\sum_{f \in F} \sum_{(i,j) \in E} d_{ij} . X_{ij}^f$$
f_3	Cost	$$\sum_{f \in F} \sum_{(i,j) \in E} w_{ij} . X_{ij}^f$$
f_4	Bandwidth consumption	$$\sum_{f \in F} \sum_{(i,j) \in E} bw_f . X_{ij}^f$$
f_5	Packet loss rate	$$\sum_{f \in F} \prod_{(i,j) \in E} plr_{ij} . X_{ij}^f$$
f_6	Blocking probability	$\min 1 - BP$, where $BP = \dfrac{Connection_{real}}{connection_{total}}$ or $$\max Connection_{real} = \sum_{f \in F} \max \left(X_{ij}^f \right)_{(i,j) \in}$$
f_7	Maximum link utilization	$\min \ \alpha$ where $$\alpha = \max \left\{ \alpha_{ij} \right\}, \text{ where } \alpha_{ij} = \frac{\sum_{f \in F} bw_f . X_{ij}^f}{c_{ij}}$$

Constraints

	Name	Expression
c_1	Source node	$$\sum_{(i,j) \in E} X_{ij}^f = 1, f \in F, i = s$$
c_2	Destination node	$$\sum_{(j,i) \in E} X_{ji}^f = -1, i = t, f \in F$$
c_3	Intermediate node	$$\sum_{(i,j) \in E} X_{ij}^f - \sum_{(j,i) \in E} X_{ji}^f = 0, f \in F, i \neq s, i \neq t$$
c_4	Link capacity	$$\sum_{f \in F} bw_f . X_{ij}^f \leq c_{ij}, (i,j) \in E$$

Table 4.4   Objective Functions

	Function	Expression
f_1	Hop count	$\min \sum\limits_{f \in F} \sum\limits_{t \in T_f} \sum\limits_{(i,j) \in E} X_{ij}^{tf}$
f_2	End-to-end delay	$\min \sum\limits_{f \in F} \sum\limits_{t \in T_f} \sum\limits_{(i,j) \in E} d_{ij}.X_{ij}^{tf}$
f_3	Cost	$\min \sum\limits_{f \in F} \sum\limits_{(i,j) \in E} w_{ij}.\max\left(X_{ij}^{tf}\right)_{t \in T_f}$
f_4	Bandwidth consumption	$\min \sum\limits_{f \in F} \sum\limits_{(i,j) \in E} bw_f.\max\left(X_{ij}^{tf}\right)_{t \in T_f}$
f_5	Blocking probability	$\min 1 - BP, \text{ where } BP = \dfrac{\text{Connection}_{real}}{\text{connection}_{total}} \quad \text{or}$ $\max Connection_{real} = \sum\limits_{f \in F} \max\left(X_{ij}^{tf}\right)_{(i,j) \in E, t \in T_f}$
f_6	Maximum link utilization	$\min \ \alpha \ \text{where}$ $\alpha = \max\left\{\alpha_{ij}\right\}, \text{ where } \alpha_{ij} = \dfrac{\sum\limits_{f \in F} bw_f.\max\left(X_{ij}^{f}\right)_{t \in T_f}}{c_{ij}}$

Other multicast functions

f_7	Hop count average	$\min \dfrac{Total\_Hop\_Count}{\sum\limits_{f \in F}\left|T_f\right|}$ $= \min \dfrac{\sum\limits_{f \in F} \sum\limits_{t \in T_f} \sum\limits_{(i,j) \in E} X_{ij}^{tf}}{\sum\limits_{f \in F}\left|T_f\right|}$
f_8	Maximal hop count	$\min \max\limits_{\substack{f \in F, \\ t \in T_f}}\left\{\sum\limits_{(i,j) \in E} X_{ij}^{tf}\right\}$

Table 4.4   Objective Functions (continued)

f_9	Maximal hop count variation	$\min \max\limits_{f \in F}\left\{H_f\right\}$ where $$H_f = \max\limits_{t \in T_f}\left\{\sum_{(i,j) \in E} X_{ij}^{tf}\right\} - \min\limits_{t \in T_f}\left\{\sum_{(i,j) \in E} X_{ij}^{tf}\right\}$$
f_{10}	Average delay	$$\min \frac{Total\_Delay}{\sum\limits_{f \in F}\left|T_f\right|} = \min \frac{\sum\limits_{f \in F}\sum\limits_{t \in T_f}\sum\limits_{(i,j) \in E} d_{ij}.X_{ij}^{tf}}{\sum\limits_{f \in F}\left|T_f\right|}$$
f_{11}	Maximal delay	$$\min \max\limits_{\substack{f \in F, \\ t \in T_f}}\left\{\sum_{(i,j) \in E} d_{ij}.X_{ij}^{tf}\right\}$$
f_{12}	Maximal delay variation	$\min \max\limits_{f \in F}\left\{\Delta_f\right\}$ where $$\Delta_f = \max\limits_{t \in T_f}\left\{\sum_{(i,j) \in E} d_{ij}.X_{ij}^{tf}\right\} - \min\limits_{t \in T_f}\left\{\sum_{(i,j) \in E} d_{ij}.X_{ij}^{tf}\right\}$$
f_{13}	Average cost	$$\min \frac{Total\_Cost}{\sum\limits_{f \in F}\left|T_f\right|} = \min \frac{\sum\limits_{f \in F}\sum\limits_{(i,j) \in E} w_{ij}.\max\left(X_{ij}^{tf}\right)_{t \in T_f}}{\sum\limits_{f \in F}\left|T_f\right|}$$
f_{14}	Maximal cost	$$\min \max\limits_{f \in F}\left\{\sum_{(i,j) \in E} \max\left(X_{ij}^{tf}\right)_{t \in T_f}\right\}$$

Constraints

	Name	*Expression*
c_1	Source node	$$\sum_{(i,j) \in E} X_{ij}^{tf} = 1,\ t \in T_f, f \in F, i = s$$

Table 4.4 Objective Functions (continued)

c_2	Destination node	$$\sum_{(j,i)\in E} X_{ji}^{tf} = -1,\, i = t, t \in T_f,\, f \in F$$
c_3	Intermediate node	$$\sum_{(i,j)\in E} X_{ij}^{tf} - \sum_{(j,i)\in E} X_{ji}^{tf} = 0,\, t \in T_f,\, f \in F, i \ne s, i \notin T_f$$
c_4	Link capacity	$$\sum_{f\in F} bw_f . \max\left(X_{ij}^{tf}\right)_{t\in T_f} \le c_{ij},\, (i,j) \in E$$

sum method from among the traditional methods that we explained in previous chapters. As in Chapter 5, we will only use metaheuristics to solve the problems discussed.

4.5.1 Unicast Transmission Using Hop Count and Delay

To understand the multi-objective optimization process, in this section we will use several traditional methods to analyze the hop count (f_1) and delay (f_2) functions in a unicast transmission in a very simple network topology (Figure 4.23). In the results, one will add the values of the bandwidth consumption function (f_4) so that these results may later be analyzed when function f_4 is also optimized.

4.5.1.1 Weighted Sum

The first traditional method we will use is the weighted sum method.

To solve the problem using this method we must rewrite the functions that we are going to optimize:

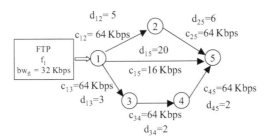

Figure 4.23 Example 1.

$$\min Z = r_1 . \sum_{f \in F} \sum_{(i,j) \in E} X_{ij}^f + r_2 . \sum_{f \in F} \sum_{(i,j) \in E} d_{ij} . X_{ij}^f \qquad (1)$$

subject to

$$\sum_{(i,j) \in E} X_{ij}^f = 1, f \in F, i = s$$

$$\sum_{(j,i) \in E} X_{ji}^f = -1, i = t, f \in F$$

$$\sum_{(i,j) \in E} X_{ij}^f - \sum_{(j,i) \in E} X_{ji}^f = 0, f \in F, i \neq s, i \neq t$$

$$\sum_{f \in F} bw_f . X_{ij}^f \leq c_{ij}, (i,j) \in E$$

$$X_{ij}^f = \{0,1\}, 0 \leq r_i \leq 1, \sum_{i=1}^{2} r_i = 1$$

To solve this problem, we can use a solver such as cplex, which can solve mixed integer programming (MIP) type problems.

Figure 4.24 shows the three possible values. Solution {(1, 5)} is not feasible due to the link capacity constraint (1.c.4), because the demand of f_1 is 32 Kbps and the capacity of the link (1, 5) is only 16 Kbps. If this constraint did not exist, path {(1, 5)} would be the best value at the hops level.

For different values r_i we can obtain the solutions in Table 4.5.

4.5.1.2 ε-Constraint

The second traditional method that we will use is ε-constraint. To solve the problem using this method we must rewrite the functions. Because in this case we have two objective functions, the model may be rewritten in two ways:

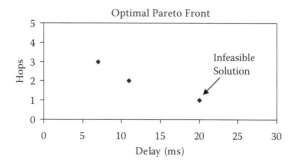

Figure 4.24   Optimal values in multi-objective optimization.

Table 4.5   Weighted Sum Solutions

Sol	r_1	r_2	f_1 (hops)	f_2 (ms)	f_4 (Kbps)	Path
1	1	0	2	11	64	{(1, 2), (2, 5)}
2	0	1	3	7	96	{(1, 3), (3, 4), (4, 5)}
3	0.5	0.5	3	7	96	{(1, 3), (3, 4), (4, 5)}
4	0.1	0.9	3	7	96	{(1, 3), (3, 4), (4, 5)}
5	0.9	0.1	2	11	64	{(1, 2), (2, 5)}
6	0.8	0.2	2	11	64	{(1, 2), (2, 5)}
7	0.7	0.3	3	7	96	{(1, 3), (3, 4), (4, 5)}

When function f_1 is an objective function and function f_2 is a constraint

When function f_2 is objective and function f_1 is a constraint

When function f_1 is an objective function, the model is

$$\min Z = \sum_{f \in F} \sum_{(i,j) \in E} X_{ij}^f \qquad (2)$$

subject to

$$\sum_{f \in F} \sum_{(i,j) \in E} d_{ij}.X_{ij}^f \le \varepsilon_2$$

$$\sum_{(i,j)\in E} X_{ij}^{f} = 1, f \in F, i = s$$

$$\sum_{(j,i)\in E} X_{ji}^{f} = -1, i = t, f \in F$$

$$\sum_{(i,j)\in E} X_{ij}^{f} - \sum_{(j,i)\in E} X_{ji}^{f} = 0, f \in F, i \neq s, i \neq t$$

$$\sum_{f\in F} bw_{f}.X_{ij}^{f} \leq c_{ij}, (i,j) \in E$$

$$X_{ij}^{f} = \{0,1\}$$

When function f_2 is an objective function, the model is

$$\min Z = \sum_{f\in F} \sum_{(i,j)\in E} d_{ij}.X_{ij}^{f} \tag{3}$$

subject to

$$\sum_{f\in F} \sum_{(i,j)\in E} X_{ij}^{f} \leq \varepsilon_{1}$$

$$\sum_{(i,j)\in E} X_{ij}^{f} = 1, f \in F, i = s$$

$$\sum_{(j,i)\in E} X_{ji}^{f} = -1, i = t, f \in F$$

$$\sum_{(i,j)\in E} X_{ij}^{f} - \sum_{(j,i)\in E} X_{ji}^{f} = 0, f \in F, i \neq s, i \neq t$$

$$\sum_{f \in F} bw_f . X_{ij}^f \le c_{ij}, (i, j) \in E$$

$$X_{ij}^f = \{0, 1\}$$

We can use cplex, which can solve MIP type problems, to solve this problem using the ε-constraint method.

Table 4.6 and 4.7 show the solutions to the model (2) when different values ε₁ and ε₂ are considered.

4.5.1.3 Weighted Metrics

The third traditional model that we will use is the weighted metrics model. To solve the problem using this method we must rewrite the functions. First, we will consider a value of $r = 2$ and then a value of $r = \infty$.

When $r = 2$, the model is

$$\min Z = \left[w_1 . \left| Z_1 - \sum_{f \in F} \sum_{(i,j) \in E} X_{ij}^f \right|^2 + w_2 . \left| Z_2 - \sum_{f \in F} \sum_{(i,j) \in E} d_{ij} . X_{ij}^f \right|^2 \right]^{1/2} \quad (4)$$

Table 4.6 ε-Constraint Solution 1

Sol	2	f_1 (hops)	f_2 (ms)	f_4 (Kbps)	Path
1	30	2	11	64	{(1, 2), (2, 5)}
2	20	2	11	64	{(1, 2), (2, 5)}
3	10	3	7	96	{(1, 3), (3, 4), (4, 5)}
4	7	3	7	96	{(1, 3), (3, 4), (4, 5)}
5	6				Solution not feasible

Table 4.7 ε-Constraint Solution 2

Sol	1	f_1 (hops)	f_2 (ms)	f_4 (Kbps)	Path
1	10	3	7	96	{(1, 3), (3, 4), (4, 5)}
2	5	3	7	96	{(1, 3), (3, 4), (4, 5)}
3	3	3	7	96	{(1, 3), (3, 4), (4, 5)}
4	2	2	11	64	{(1, 2), (2, 5)}
5	1				Solution not feasible

subject to

$$\sum_{(i,j)\in E} X_{ij}^f = 1, f \in F, i = s$$

$$\sum_{(j,i)\in E} X_{ji}^f = -1, i = t, f \in F$$

$$\sum_{(i,j)\in E} X_{ij}^f - \sum_{(j,i)\in E} X_{ji}^f = 0, f \in F, i \neq s, i \neq t$$

$$\sum_{f\in F} bw_f . X_{ij}^f \leq c_{ij}, (i,j) \in E$$

$$X_{ij}^f = \{0,1\}$$

$$1 \leq r < \infty$$

$$0 \leq w_i \leq 1, i = \{1, ..., 2\}$$

$$\sum_{i=1}^{2} w_i = 1$$

and when $r = \infty$, the model is rewritten in the following way:

$$\min Z = \max \left[w_1 . \left| Z_1 - \sum_{f\in F} \sum_{(i,j)\in E} X_{ij}^f \right|, w_2 . \left| Z_2 - \sum_{f\in F} \sum_{(i,j)\in E} d_{ij} . X_{ij}^f \right| \right] \quad (5)$$

subject to

$$\sum_{(i,j)\in E} X_{ij}^f = 1, f \in F, i = s$$

$$\sum_{(j,i)\in E} X_{ji}^{f} = -1, i = t, f \in F$$

$$\sum_{(i,j)\in E} X_{ij}^{f} - \sum_{(j,i)\in E} X_{ji}^{f} = 0, f \in F, i \neq s, i \neq t$$

$$\sum_{f\in F} bw_{f}.X_{ij}^{f} \leq c_{ij}, (i,j) \in E$$

$$X_{ij}^{f} = \{0,1\}$$

$$0 \leq w_{i} \leq 1, i = \{1, \ldots, 2\}$$

$$\sum_{i=1}^{2} w_{i} = 1$$

To solve this method, we can use a solver such as the standard branch and bound (SBB), which can solve mixed integer nonlinear programming (MINLP) type problems, because we include the absolute value function ($|x|$).

Table 4.8 shows the solutions to model (4) when $r = 2$, and different values Z_1, Z_2, w_1, and w_2 are considered.

For $r = \infty$ one obtains exactly the same results.

Table 4.8 Weighted Metric Solution with r=2 and r=∞

Sol	w_1	w_2	Z_1	Z_2	f_1 (hops)	f_2 (ms)	f_4 (Kbps)	Path
1	0.9	0.1	5	20	3	7	96	{(1, 3), (3, 4), (4, 5)}
2	0.9	0.1	1	20	2	11	64	{(1, 2), (2, 5)}
3	0.8	0.2	1	20	2	11	64	{(1, 2), (2, 5)}
4	0.1	0.9	2	20	2	11	64	{(1, 2), (2, 5)}
5	0.1	0.9	3	20	2	11	64	{(1, 2), (2, 5)}
6	0.5	0.5	3	3	3	7	96	{(1, 3), (3, 4), (4, 5)}
7	0.9	0.1	3	3	3	7	96	{(1, 3), (3, 4), (4, 5)}
8	0.1	0.9	3	3	3	7	96	{(1, 3), (3, 4), (4, 5)}
9	0.9	0.1	1	1	2	11	64	{(1, 2), (2, 5)}
10	0.5	0.5	1	1	3	7	96	{(1, 3), (3, 4), (4, 5)}
11	0.1	0.9	1	1	3	7	96	{(1, 3), (3, 4), (4, 5)}

4.5.1.4 Benson Method

The last traditional method that we will use to solve the multi-objective problem is the Benson method. To solve the problem using this method, we must rewrite the functions in the following way:

$$
\max Z = \max\left(0, Z_1 - \sum_{f \in F}\sum_{(i,j) \in E} X_{ij}^f\right) + \max\left(0, Z_2 - \sum_{f \in F}\sum_{(i,j) \in E} d_{ij}.X_{ij}^f\right)
$$

(6)

subject to

$$
\sum_{f \in F}\sum_{(i,j) \in E} X_{ij}^f \le Z_1
$$

$$
\sum_{f \in F}\sum_{(i,j) \in E} d_{ij}.X_{ij}^f \le Z_2
$$

$$
\sum_{(i,j) \in E} X_{ij}^f = 1, f \in F, i = s
$$

$$
\sum_{(j,i) \in E} X_{ji}^f = -1, i = t, f \in F
$$

$$
\sum_{(i,j) \in E} X_{ij}^f - \sum_{(j,i) \in E} X_{ji}^f = 0, f \in F, i \ne s, i \ne t
$$

$$
\sum_{f \in F} bw_f.X_{ij}^f \le c_{ij}, (i,j) \in E
$$

$$
X_{ij}^f = \{0,1\}
$$

Table 4.9 shows the solutions to the model for the values Z_1 and Z_2, for which a solution was found.

Table 4.9   Benson Solutions

Sol	Z_1	Z_2	f_1 (hops)	f_2 (ms)	f_4 (Kbps)	Path
1	5	20	2	11	64	{(1, 2), (2, 5)}
2	2	20	2	11	64	{(1, 2), (2, 5)}
3	3	20	2	11	64	{(1, 2), (2, 5)}
4	3	7	3	7	96	{(1, 3), (3, 4), (4, 5)}
5	4	10	3	7	96	{(1, 3), (3, 4), (4, 5)}

4.5.2 Multicast Transmission Using Hop Count and Delay

As we did in the previous section for unicast transmission, we will use several traditional methods to analyze the hop count (f_1) and delay (f_2) functions in a multicast transmission in a very simple network topology (Figure 4.25).

4.5.2.1 Weighted Sum

To solve the problem with this method, we must rewrite the functions in the following way:

$$\min Z = r_1 . \sum_{f \in F} \sum_{i \in T_f} \sum_{(i,j) \in E} X_{ij}^{tf} + r_2 . \sum_{f \in F} \sum_{i \in T_f} \sum_{(i,j) \in E} d_{ij} . X_{ij}^{tf} \tag{7}$$

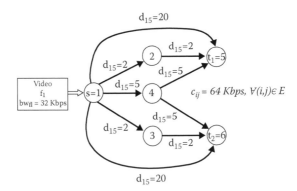

Figure 4.25   Example 1.

subject to

$$\sum_{(i,j)\in E} X_{ij}^{tf} = 1, t \in T_f, f \in F, i = s$$

$$\sum_{(j,i)\in E} X_{ji}^{tf} = -1, i = t, t \in T_f, f \in F$$

$$\sum_{(i,j)\in E} X_{ij}^{tf} - \sum_{(j,i)\in E} X_{ji}^{tf} = 0, t \in T_f, f \in F, i \neq s, i \neq t$$

$$\sum_{f\in F} bw_f . \max\left(X_{ij}^{tf}\right)_{t\in T_f} \leq c_{ij}, (i,j) \in E \qquad (7.c.4)$$

$$X_{ij}^{tf} = \{0,1\}, 0 \leq r_i \leq 1, \sum_{i=1}^{2} r_i = 1$$

If we remove the link capacity constraint (7.c.4), the problem is of the mixed integer programming (MIP) type and we can use the cplex as solver. Considering the link capacity constraint, because a max function is included, we have a DNLP (nonlinear programming with discontinuous derivatives) type problem, and in this case, one must use solvers such as Sparse Nonlinear Optimizer (SNOPT) or Constraint Nonlinear Optimizer (COMOPT).

Table 4.10 shows these solutions.

Table 4.10   Optimal and Feasible Solutions

Sol	f_1 (hops)	f_2 (ms)	f_4 (Kbps)	Tree
1	3	30	96	[{(1, 5)}, {(1, 4), (4, 6)}]
2	3	30	96	[{(1, 4), (4, 5)}, {(1, 6)}]
3	4	20	96	[{(1, 4), (4, 5)}, {(1, 4), (4, 6)}]
4	4	14	128	[{(1, 2), (2, 5)}, {(1, 4), (4, 6)}]
5	4	14	128	[{(1, 4), (4, 5}, {(1, 3), (3, 6)}]
6	2	40	64	[{(1, 5)}, {(1, 6)}]
7	4	8	128	[{(1, 2), (2, 5)}, {(1, 3), (3, 6)}]
8	3	24	96	[{(1, 5)}, {(1, 3), (3, 6)}]
9	3	24	96	[{(1, 2), (2, 5)}, {(1, 6)}]

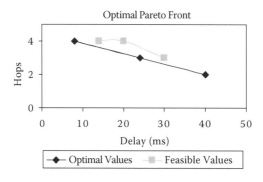

Figure 4.26 Optimal values in multi-objective optimization.

Table 4.11 Weighted Sum Solutions

Sol	r_1	r_2	f_1 (hops)	f_2 (ms)	f_4 (Kbps)	Tree
1	1	0	2	40	64	[{(1, 5)}, {(1, 6)}]
2	0	1	4	8	128	[{(1, 2), (2, 5)}, {(1, 3), (3, 6)}]
3	1	0	3	24	96	[{(1, 5)}, {(1, 3), (3, 6)}]

Figure 4.26 shows these feasible solutions. Because for some solutions (1 and 2, 4 and 5, and 8 and 9) f_1 and f_2 coincide, the network's administrator has to select one of them.

If we execute the model for different values r_i, we can obtain the solutions in Table 4.11.

In this case, if there is more than one solution that gives the same values for the optimization function, it may happen that only one of these solutions is found. For example, as we saw in the previous case, trees [{(1, 2), (2, 5)}, {(1, 6)}] and [{(1, 5)}, {(1, 3), (3, 6)}] have exactly the same values for hops (3) and delay (24), but in this case, only one of the trees has been found.

4.5.2.2 ε–Constraint

As we have already mentioned in the unicast transmission case, this model may be rewritten in two ways:

When function f_1 is an objective function and function f_2 is a constraint

When function f_2 is an objective function and function f_1 is a constraint

When function f_1 is the objective function, the model is

$$\min Z = \sum_{f \in F} \sum_{t \in T_f} \sum_{(i,j) \in E} X_{ij}^{tf} \tag{8}$$

subject to

$$\sum_{f \in F} \sum_{t \in T_f} \sum_{(i,j) \in E} d_{ij} . X_{ij}^{tf} \leq \varepsilon_2$$

$$\sum_{(i,j) \in E} X_{ij}^{tf} = 1, t \in T_f, f \in F, i = s$$

$$\sum_{(j,i) \in E} X_{ji}^{tf} = -1, i = t, t \in T_f, f \in F$$

$$\sum_{(i,j) \in E} X_{ij}^{tf} - \sum_{(j,i) \in E} X_{ji}^{tf} = 0, t \in T_f, f \in F, i \neq s, i \neq t$$

$$\sum_{f \in F} bw_f . \max\left(X_{ij}^{tf}\right)_{t \in T_f} \leq c_{ij}, (i,j) \in E$$

$$X_{ij}^{tf} = \left\{0,1\right\}$$

When function f_2 is the objective function, the model is

$$\min Z = \sum_{f \in F} \sum_{t \in T_f} \sum_{(i,j) \in E} d_{ij} . X_{ij}^{tf} \tag{9}$$

subject to

$$\sum_{f \in F} \sum_{t \in T_f} \sum_{(i,j) \in E} X_{ij}^{tf} \leq \varepsilon_1$$

$$\sum_{(i,j)\in E} X_{ij}^{tf} = 1, t \in T_f, f \in F, i = s$$

$$\sum_{(j,i)\in E} X_{ji}^{tf} = -1, i = t, t \in T_f, f \in F$$

$$\sum_{(i,j)\in E} X_{ij}^{tf} - \sum_{(j,i)\in E} X_{ji}^{tf} = 0, t \in T_f, f \in F, i \neq s, i \neq t$$

$$\sum_{f \in F} bw_f \cdot \max\left(X_{ij}^{tf}\right)_{t\in T_f} \leq c_{ij}, (i,j) \in E$$

$$X_{ij}^{tf} = \{0,1\}$$

As we have already mentioned, if we remove the link capacity (7.c.4) constraint, the problem is of the MIP type, and if we consider the link capacity constraint, we have a DNLP type problem.

When f_1 and f_2 are the objective function, for different values ε_2 we can obtain the solutions in Table 4.11 and 4.12.

We can see that when we optimize f_1 we find tree [{(1, 2), (2, 5)}, {(1, 6)}], and when we optimize f_2 we find tree [{(1, 5)}, {(1, 3), (3, 6)}], where the values of the functions are the same. This shows us that if we optimize all functions, we can find more solutions than when we used the weighted sum method, where only one of these trees was found.

Table 4.11   ε-Constraint Solution

Sol	2	f_1 (hops)	f_2 (ms)	f_4 (Kbps)	Tree
1	50	2	40	64	[{(1, 5)}, {(1, 6)}]
2	30	3	24	96	[{(1, 2), (2, 5)}, {(1, 6)}]
3	20	4	8	128	[{(1, 2), (2, 5)}, {(1, 3), (3, 6)}]

Table 4.12   ε-Constraint Solution

Sol	1	f_1 (hops)	f_2 (ms)	f_4 (Kbps)	Tree
1	10	4	8	128	[{(1, 2), (2, 5)}, {(1, 3), (3, 6)}]
2	3	3	24	96	[{(1, 5)}, {(1, 3), (3, 6)}]
3	2	2	40	64	[{(1, 5)}, {(1, 6)}]

4.5.2.3 Weighted Metrics

To solve the problem with this method, we must rewrite the functions. First, we will consider a value of $r = 2$ and then a value of $r = \infty$.

When $r = 2$, the model is

$$
\min Z = \left[w_1 . \left| Z_1 - \sum_{f \in F} \sum_{t \in T_f} \sum_{(i,j) \in E} X_{ij}^{tf} \right|^2 + w_2 . \left| Z_2 - \sum_{f \in F} \sum_{t \in T_f} \sum_{(i,j) \in E} d_{ij} . X_{ij}^{tf} \right|^2 \right]^{1/2}
$$

(10)

subject to

$$
\sum_{(i,j) \in E} X_{ij}^{tf} = 1, t \in T_f, f \in F, i = s
$$

$$
\sum_{(j,i) \in E} X_{ji}^{tf} = -1, i = t, t \in T_f, f \in F
$$

$$
\sum_{(i,j) \in E} X_{ij}^{tf} - \sum_{(j,i) \in E} X_{ji}^{tf} = 0, t \in T_f, f \in F, i \neq s, i \neq t
$$

$$
\sum_{f \in F} bw_f . \max \left(X_{ij}^{tf} \right)_{t \in T_f} \leq c_{ij}, (i,j) \in E
$$

$$
X_{ij}^{tf} = \left\{ 0, 1 \right\}
$$

$$1 \leq r < \infty$$

$$0 \leq w_i \leq 1, \, i = \{1, \ldots, 2\}$$

$$\sum_{i=1}^{2} w_i = 1$$

When $r = \infty$, the model is

$$\min Z = \max \left[w_1 . \left| Z_1 - \sum_{f \in F} \sum_{t \in T_f} \sum_{(i,j) \in E} X_{ij}^{tf} \right|, w_2 . \left| Z_2 - \sum_{f \in F} \sum_{t \in T_f} \sum_{(i,j) \in E} d_{ij} . X_{ij}^{tf} \right| \right]$$

(11)

subject to

$$\sum_{(i,j) \in E} X_{ij}^{tf} = 1, t \in T_f, f \in F, i = s$$

$$\sum_{(j,i) \in E} X_{ji}^{tf} = -1, i = t, t \in T_f, f \in F$$

$$\sum_{(i,j) \in E} X_{ij}^{tf} - \sum_{(j,i) \in E} X_{ji}^{tf} = 0, t \in T_f, f \in F, i \neq s, i \neq t$$

$$\sum_{f \in F} bw_f . \max \left(X_{ij}^{tf} \right)_{t \in T_f} \leq c_{ij}, (i,j) \in E$$

$$X_{ij}^{tf} = \{0,1\}$$

$$0 \leq w_i \leq 1, \, i = \{1, \ldots, 2\}$$

$$\sum_{i=1}^{2} w_i = 1$$

As happened with unicast transmission, this is a mixed integer nonlinear programming (MINLP) type problem because we include the absolute value function ($|x|$), and to solve it, we must use solver SBB.

Table 4.13 and 4.14 shows the solutions to the model (10) when $r = 2$ and $r = \infty$ one considers different values of Z_1, Z_2, w_1, and w_2.

It may happen that certain values that have been obtained with $r = 2$, for example, solution 3, cannot be easily obtained when $r = \infty$, or even when other methods are used.

4.5.2.4 Benson Method

As we did for the unicast transmission, the last traditional model that we will use to solve the multi-objective problem is the Benson method.

To solve the problem using this method we must rewrite the functions:

$$
\max Z = \max\left(0, Z_1 - \sum_{f \in F} \sum_{i \in T_f} \sum_{(i,j) \in E} X_{ij}^{tf} \right) + \max\left(0, Z_2 - \sum_{f \in F} \sum_{i \in T_f} \sum_{(i,j) \in E} d_{ij}.X_{ij}^{tf} \right)
$$

$$(12)$$

Table 4.13 Weighted Metric Solutions with r=2

Sol	w_1	w_2	Z_1	Z_2	f_1 (hops)	f_2 (ms)	f_4 (Kbps)	Tree
1	0.1	0.9	5	50	2	40	64	[{(1, 5)}, {(1, 6)}]
2	0.9	0.1	5	50	2	40	64	[{(1, 5)}, {(1, 6)}]
3	0.1	0.9	5	25	3	24	96	[{(1, 5)}, {(1, 3), (3, 6)}]
4	0.1	0.9	5	10	4	8	128	[{(1, 2), (2, 5)}, {(1, 3), (3, 6)}]

Table 4.14 Weighted Metric Solutions with r=∞

Sol	w_1	w_2	Z_1	Z_2	f_1 (hops)	f_2 (ms)	f_4 (Kbps)	Tree
1	0.1	0.9	5	50	2	40	64	[{(1, 5)}, {(1, 6)}]
2	0.9	0.1	5	50	2	40	64	[{(1, 5)}, {(1, 6)}]
3	0.1	0.9	5	10	4	8	128	[{(1, 2), (2, 5)}, {(1, 3), (3, 6)}]

subject to

$$\sum_{f \in F} \sum_{t \in T_f} \sum_{(i,j) \in E} X_{ij}^{tf} \le Z_1$$

$$\sum_{f \in F} \sum_{t \in T_f} \sum_{(i,j) \in E} d_{ij}.X_{ij}^{tf} \le Z_2$$

$$\sum_{(i,j) \in E} X_{ij}^{tf} = 1, t \in T_f, f \in F, i = s$$

$$\sum_{(j,i) \in E} X_{ji}^{tf} = -1, i = t, t \in T_f, f \in F$$

$$\sum_{(i,j) \in E} X_{ij}^{tf} - \sum_{(j,i) \in E} X_{ji}^{tf} = 0, t \in T_f, f \in F, i \ne s, i \ne t$$

$$\sum_{f \in F} bw_f.\max\left(X_{ij}^{tf}\right)_{t \in T_f} \le c_{ij}, (i, j) \in E$$

$$X_{ij}^{tf} = \{0, 1\}$$

Table 4.15 shows the solutions to the model (12) for different values of Z_1 and Z_2.

Table 4.15 Benson Method Solutions

Sol	Z_1	Z_2	f_1 (hops)	f_2 (ms)	f_4 (Kbps)	Tree
1	5	50	4	8	128	[{(1, 2), (2, 5)}, {(1, 3), (3, 6)}]
2	5	25	4	8	128	[{(1, 2), (2, 5)}, {(1, 3), (3, 6)}]
3	3	24	3	24	96	[{(1, 2), (2, 5)}, {(1, 6)}]
4	2	50	2	40	64	[{(1, 5)}, {(1, 6)}]

4.5.3 Unicast Transmission Using Hop Count, Delay, and Bandwidth Consumption

For this case, we will use the functions hop count (f_1), delay (f_2), and bandwidth consumption (f_4) in a unicast transmission.

In this section we will consider the weighted sum traditional method to subsequently be able to compare the results with a metaheuristic.

To solve the problem using this method, we must rewrite the functions in the following way:

$$\min Z = r_1 . \sum_{f \in F} \sum_{(i,j) \in E} X_{ij}^f + r_2 . \sum_{f \in F} \sum_{(i,j) \in E} d_{ij} . X_{ij}^f + r_3 . \sum_{f \in F} \sum_{(i,j) \in E} bw_f . X_{ij}^f$$

(13)

subject to

$$\sum_{(i,j) \in E} X_{ij}^f = 1, f \in F, i = s$$

$$\sum_{(j,i) \in E} X_{ji}^f = -1, i = t, f \in F$$

$$\sum_{(i,j) \in E} X_{ij}^f - \sum_{(j,i) \in E} X_{ji}^f = 0, f \in F, i \neq s, i \neq t$$

$$\sum_{f \in F} bw_f . X_{ij}^f \leq c_{ij}, (i,j) \in E$$

$$X_{ij}^f = \{0,1\}, 0 \leq r_i \leq 1, \sum_{i=1}^{3} r_i = 1$$

The problem is a MIP type problem, and therefore, we can use Cplex to solve it.

Figure 4.27 shows the topology considered. In this case, the links are T1, and therefore, the capacity considered is 1536 Kbps.

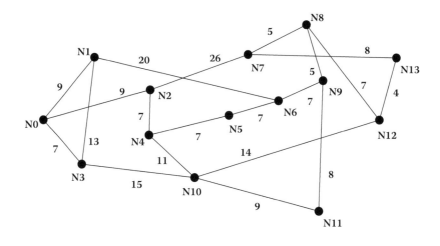

Figure 4.27 Example 2 network.

Example 1

Table 4.16 shows the solutions for different values r_i when we transmit a single 512-Kbps flow f_1 from node 0 to node 12.

In this table we can see that notwithstanding the values of r_i, there is only one path that simultaneously optimizes the three functions.

Example 2

Table 4.17 shows the solutions for different values r_i when we transmit a single 512-Kbps flow f_1 from node 0 to node 13.

Table 4.16 Example 1 Solutions

Sol	r_1	r_2	r_3	f_1 (hops)	f_2 (ms)	f_4 (Kbps)	Path
1	0	0	1	3	36	1536	{(0, 3), (3, 10), (10, 12)}
2	0	1	0	3	36	1536	{(0, 3), (3, 10), (10, 12)}
3	1	0	0	3	36	1536	{(0, 3), (3, 10), (10, 12)}
4	0.3	0.3	0.4	3	36	1536	{(0, 3), (3, 10), (10, 12)}

Table 4.17   Example 2 Solutions

Sol	r_1	r_2	r_3	f_1 (hops)	f_2 (ms)	f_4 (Kbps)	Path
1	0	0	1	3	43	1536	{(0, 2), (2, 7), (7, 13)}
2	0	1	0	4	40	2048	{(0, 3), (3, 10), (10, 12), (12, 13)}
3	1	0	0	3	43	1536	{(0, 2), (2, 7), (7, 13)}
4	0.3	0.3	0.4	3	43	1536	{(0, 2), (2, 7), (7, 13)}

Example 3

Table 4.18 shows the solutions for different values r_i when we transmit three flows, f_1, f_2, and f_3, of 512 Kbps from node 0 to node 13.

Table 4.18   Example 3 Solutions

Sol	r_1	r_2	r_3	f_1 (hops)	f_2 (ms)	f_4 (Kbps)	Path
1	0	0	1	9	129	4608	{(0, 2), (2, 7), (7, 13)}
2	0	1	0	12	120	6144	{(0, 3), (3, 10), (10, 12), (12, 13)}
3	1	0	0	9	129	4608	{(0, 2), (2, 7), (7, 13)}
4	0.3	0.3	0.4	9	129	4608	{(0, 2), (2, 7), (7, 13)}

Example 4

Table 4.19 shows the solutions for different values r_i when we transmit two 1024-Kbps flows f_1 and f_2 from node 0 to node 13.

Table 4.19   Example 4 Solutions

Sol	r_1	r_2	r_3	f_1 (hops)	f_2 (ms)	f_4 (Kbps)	Flow Path
1	0	0	1	7	83	7168	f_1 {(0, 3), (3, 10), (10, 12), (12, 13)} f_2 {(0, 2), (2, 7), (7, 13)}
2	0	1	0	7	83	7168	f_1 {(0, 3), (3, 10), (10, 12), (12, 13)} f_2 {(0, 2), (2, 7), (7, 13)}
3	1	0	0	7	83	7168	f_1 {(0, 3), (3, 10), (10, 12), (12, 13)} f_2 {(0, 2), (2, 7), (7, 13)}
4	0.3	0.3	0.4	7	83	7168	f_1 {(0, 3), (3, 10), (10, 12), (12, 13)} f_2 {(0, 2), (2, 7), (7, 13)}

Example 5

Lastly, if in the previous example we consider an additional third flow f_3 with the same transmission rate of 1024 Kbps, any solution that one obtains is nonfeasible due to the link capacity constraint. Using the maximum link utilization function, the flows can be fractioned through the different paths and it would be possible to find feasible solutions. In this case, as we have seen previously, the variables $X_{ij}^f = [0,1]$.

4.5.4 Multicast Transmission Using Hop Count, Delay, and Bandwidth Consumption

In this section we will do the analysis using the functions hop count, delay, and bandwidth consumption in a multicast transmission over the topology of Figure 4.27. As we did in the previous section, we will use the weighted sum traditional method and then compare the results with a metaheuristic.

To solve the problem with this method, we must rewrite the functions in the following way:

$$\min Z = r_1 . \sum_{f \in F} \sum_{t \in T_f} \sum_{(i,j) \in E} X_{ij}^{tf} + r_2 . \sum_{f \in F} \sum_{t \in T_f} \sum_{(i,j) \in E} d_{ij} . X_{ij}^{tf}$$

$$+ r_3 . \sum_{f \in F} \sum_{(i,j) \in E} bw_f . \max\left(X_{ij}^{tf} \right)_{t \in T_f} \tag{14}$$

subject to

$$\sum_{(i,j) \in E} X_{ij}^{tf} = 1, t \in T_f, f \in F, i = s$$

$$\sum_{(j,i) \in E} X_{ji}^{tf} = -1, i = t, t \in T_f, f \in F$$

$$\sum_{(i,j) \in E} X_{ij}^{tf} - \sum_{(j,i) \in E} X_{ji}^{tf} = 0, t \in T_f, f \in F, i \neq s, i \neq t$$

$$\sum_{f \in F} bw_f . \max \left(X_{ij}^{tf} \right)_{t \in T_f} \le c_{ij}, (i, j) \in E$$

$$X_{ij}^{tf} = \left\{ 0, 1 \right\}, 0 \le r_i \le 1, \sum_{i=1}^{3} r_i = 1$$

Example 1

We will transmit a 512-Kbps single multicast flow f_1 from node 0 to nodes 8 and 12.

In this case we will need a solver DNLP (nonlinear programming with discontinuous derivatives) because the bandwidth consumption function uses the max function.

Table 4.20 shows the solutions for different values r_i.

In solutions 1 and 2 we see that only one function has been minimized, hop count and delay, respectively, and in both cases, the resulting tree is the same (Figure 4.28). In this case, SNOPT was used as the solver.

In solution 3 only the bandwidth function has been optimized (Figure 4.29). The resulting tree is a different tree that tried to use the same links for both destinations and, in this way,

Table 4.20   Example 1 Solution

Sol	r_1	r_2	r_3	f_1 (hops)	f_2 (ms)	f_4 (Kbps)	Tree
1	1	0	0	6	76	3072	[{(0, 2), (2, 7), (7, 8)}, {(0, 3), (3, 10), (10, 12)}]
2	0	1	0	6	76	3072	[{(0, 2), (2, 7), (7, 8)}, {(0, 3), (3, 10), (10, 12)}]
3	0	0	1	7	79	2048	[{(0, 3), (3, 10), (10, 12), (12, 8)}, {(0, 3), (3, 10), (10, 12)}]
4	0.5	0.49	0.01	7	79	2048	[{(0, 3), (3, 10), (10, 12), (12, 8)}, {(0, 3), (3, 10), (10, 12)}]
5	0.5	0.5	0	6	76	3072	[{(0, 2), (2, 7), (7, 8)}, {(0, 3), (3, 10), (10, 12)}]

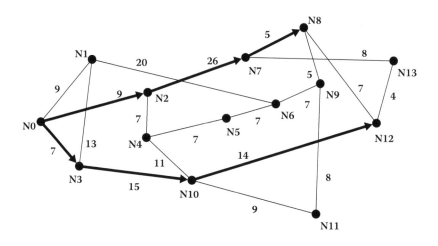

Figure 4.28   Minimizing hop count or delay.

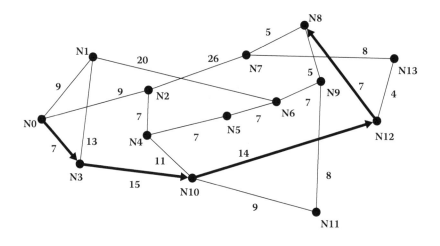

Figure 4.29   Minimizing bandwidth consumption.

minimize the bandwidth consumption used. CONOPT was used as the solver in this case.

In solution 4 we can see that even though the value of the weight associated with the bandwidth consumption function is low (r_3 = 0.01), the result still slants toward that function.

Lastly, in solution 5 we see that if we remove the bandwidth consumption function, the optimization process will try to find

Table 4.21 Example 2 Solution

Sol	r_1	r_2	r_3	f_1 (hops)	f_2 (ms)	f_4 (Kbps)	Tree
1	0	1	0	18	228	9216	[{(0, 2), (2, 7), (7, 8)}, {(0, 3), (3, 10), (10, 12)}]

the tree with the minimum value of hop count and delay, and the solution corresponds to the one previously found in 1 and 2.

Example 2

We will transmit three 512-Kbps multicast flows, f_1, f_2, and f_3, each of them from origin node 0 to destination nodes 8 and 12.

Table 4.21 shows the only solution that was found in the tests performed with different values r_i.

4.5.5 Unicast Transmission Using Hop Count, Delay, Bandwidth Consumption, and Maximum Link Utilization

In this section we will use the hop count (f_1), delay (f_2), bandwidth consumption (f_4), and maximum link utilization (f_7) functions in a unicast transmission to find a feasible solution for the case analyzed in the latter part of Section 4.5.3, where there was no feasible solution to the problem due to the capacity of the links. By adding the maximum link utilization function we are using a technique known as load balancing, through which each flow f may be divided and transmitted through different paths. In this case, the value of variables X_{ij}^f passes from binary to real between 0 and 1. If we use the maximum link utilization function with real variables X_{ij}^f, the hop count (f_1) and delay (f_2) functions must be redefined. We must define a new vector of variables Y_{ij}^f in the following way:

$$Y_{ij}^f = \left\lceil X_{ij}^f \right\rceil = \begin{cases} 0, X_{ij}^f = 0 \\ 1, 0 < X_{ij}^f \leq 1 \end{cases}$$

where Y_{ij}^f denotes whether link (i, j) is used (1) or not (0) for the unicast transmission of flow f.

Then, the hop count function would be redefined as

$$\sum_{f \in F} \sum_{(i,j) \in E} Y_{ij}^{f}$$

and the delay function as

$$\sum_{f \in F} \sum_{(i,j) \in E} d_{ij}.Y_{ij}^{f}$$

To solve the problem with this method, we must rewrite the functions in the following way:

$$\min Z = r_1.\sum_{f \in F} \sum_{(i,j) \in E} Y_{ij}^{f} + r_2.\sum_{f \in F} \sum_{(i,j) \in E} d_{ij}.Y_{ij}^{f} + r_3.\sum_{f \in F} \sum_{(i,j) \in E} bw_f.X_{ij}^{f} + r_4.\alpha$$

(15)

subject to

$$\alpha = \max\left\{\alpha_{ij}\right\}, where \ \alpha_{ij} = \frac{\sum_{f \in F} bw_f.X_{ij}^{f}}{c_{ij}}$$

$$\sum_{(i,j) \in E} X_{ij}^{f} = 1, f \in F, i = s$$

$$\sum_{(j,i) \in E} X_{ji}^{f} = -1, i = t, f \in F$$

$$\sum_{(i,j) \in E} X_{ij}^{f} - \sum_{(j,i) \in E} X_{ji}^{f} = 0, f \in F, i \neq s, i \neq t$$

$$\sum_{f \in F} bw_f.X_{ij}^{f} \leq c_{ij}, (i,j) \in E$$

$$Y_{ij}^{f} = \left\lceil X_{ij}^{f} \right\rceil = \begin{cases} 0, X_{ij}^{f} = 0 \\ 1, 0 < X_{ij}^{f} \leq 1 \end{cases}$$

$$X_{ij}^{f} = \left\{ 0,1 \right\}, 0 \leq r_{i} \leq 1, \sum_{i=1}^{4} r_{i} = 1$$

Example

Between origin node 0 and destination node 13 we will transmit three flows, f_1, f_2, and f_3, with a transmission rate of 1024 Kbps each.

Table 4.22 shows the solutions obtained for different values r_i.

Table 4.22   Optical Solutions in Unicast

Sol	$r_1, r_2,$ r_3, r_4	f_1 (hops)	f_2 (ms)	f_4 (Kbps)	f_7 (%)	Flow Fraction Path
1	$r_4 = 1$	14	166	10752	100	f_1 100% {(0, 3), (3, 10), (10, 12), (12, 13)} f_2 50% {(0, 2), (2, 7), (7, 13)} f_2 50% {(0, 3), (3, 10), (10, 12), (12, 13)} f_3 100% {(0, 2), (2, 7), (7, 13)}
2	$r_3 = 1$	14	166	10752	100	f_1 100% {(0, 3), (3, 10), (10, 12), (12, 13)} f_2 50% {(0, 2), (2, 7), (7, 13)} f_2 50% {(0, 3), (3, 10), (10, 12), (12, 13)} f_3 100% {(0, 2), (2, 7), (7, 13)}
3	$r_2 = 1$	16	178	11776	100	f_1 100% {(0, 2), (2, 7), (7, 13)} f_2 50% {(0, 1), (1, 6), (6, 9), (9, 8), (9, 8), (8, 12), (12, 13)} f_2 50% {(0, 2), (2, 7), (7, 13)} f_3 100% {(0, 3), (3, 10), (10, 12), (12, 13)}
4	$r_1 = 1$	16	178	11776	100	f_1 100% {(0, 2), (2, 7), (7, 13)} f_2 50% {(0, 1), (1, 6), (6, 9), (9, 8), (9, 8), (8, 12), (12, 13)} f_2 50% {(0, 2), (2, 7), (7, 13)} f_3 100% {(0, 3), (3, 10), (10, 12), (12, 13)}

As we can see, with the maximum link utilization function we can find feasible solutions. In this case, the unicast flow must be divided through several paths.

This mathematical model can better be expressed using the subflow concept. In this case, the set of flows is denoted by F. Every flow $f \varepsilon F$ can be divided into K_f subflows that, once normalized, can be denoted as f_k, where $k = 1, \ldots, |K_f|$. Under this denomination, f_k shows the fraction of flow $f \varepsilon F$ that is carried by subflow k. For each flow $f \varepsilon F$ we have an origin node $s_f \varepsilon N$ and a destination node t_f. It is necessary to include the constraint $\sum_{k=1}^{|k_f|} f_k = 1$ to normalize the values f_k.

In this new mathematical formula the vector of variables X_{ij}^{f} is replaced by the new denotation $X_{ij}^{f_k}$. $X_{ij}^{f_k}$ denotes a fraction of the flow f that is transmitted by subflow k through link (i, j). Similarly, one redefines $Y_{ij}^{f_k}$,

$$Y_{ij}^{f_k} = \left\lceil X_{ij}^{f_k} \right\rceil = \begin{cases} 0, X_{ij}^{f_k} = 0 \\ 1, 0 < X_{ij}^{f_k} \le 1 \end{cases}$$

To solve the problem with this method, we must rewrite the functions in the following way adding the new index k:

$$\min Z = r_1 . \sum_{f \in F} \sum_{k \in K_f} \sum_{(i,j) \in E} Y_{ij}^{f_k} + r_2 . \sum_{f \in F} \sum_{k \in K_f} \sum_{(i,j) \in E} d_{ij} . Y_{ij}^{f_k}$$

$$+ r_3 . \sum_{f \in F} \sum_{k \in K_f} \sum_{(i,j) \in E} bw_f . X_{ij}^{f_k} + r_4 . \alpha$$

(16)

subject to

$$\alpha = \max \left\{ \alpha_{ij} \right\}, \text{ where } \alpha_{ij} = \frac{\sum_{f \in F} \sum_{k \in K_f} bw_f . X_{ij}^{f_k}}{c_{ij}}$$

$$\sum_{(i,j)\in E}\sum_{k\in K_f} X_{ij}^{f_k} = 1, f\in F, i = s_f$$

$$\sum_{(j,i)\in E}\sum_{k\in K_f} X_{ji}^{f_k} = -1, i = t_f, f\in F$$

$$\sum_{(i,j)\in E} X_{ij}^{f_k} - \sum_{(j,i)\in E} X_{ji}^{f_k} = 0, f\in F, k\in K_f, i\neq s_f, i\neq t_f$$

$$\sum_{f\in F}\sum_{k\in K_f} bw_f.X_{ij}^{f_k} \leq c_{ij}, (i,j)\in E$$

$$Y_{ij}^{f_k} = \left\lceil X_{ij}^{f_k} \right\rceil = \begin{cases} 0, X_{ij}^{f_k} = 0 \\ 1, 0 < X_{ij}^{f_k} \leq 1 \end{cases}$$

$$X_{ij}^{f_k} = \{0,1\}, 0 \leq r_i \leq 1, \sum_{i=1}^{4} r_i = 1$$

This new mathematical model facilitates denotation of the flow of traffic that is transmitting every one of the subflows. In addition, it will allow for each of these subflows to be an LSP path in the MPLS technology.

4.5.6 Multicast Transmission Using Hop Count, Delay, Bandwidth Consumption, and Maximum Link Utilization

In this case, we will use the hop count (f_1), delay (f_2), bandwidth consumption (f_4), and maximum link utilization (f_7) functions in a multicast transmission. Variables X_{ij}^{tf} also pass from binary to real between 0 and 1, and every multicast flow f can be divided and transmitted through different trees. If we use the maximum link utilization function with these real variables X_{ij}^{tf}, the hop count and delay functions must be redefined.

Let us define a new vector of variables Y_{ij}^{tf} :

$$Y_{ij}^{tf} = \left\lceil X_{ij}^{tf} \right\rceil = \begin{cases} 0, X_{ij}^{tf} = 0 \\ 1, 0 < X_{ij}^{tf} \leq 1 \end{cases}$$

that represent whether link (i, j) is used (1) or not (0) for the multicast transmission of flow f.

Due to the above, the hop count function would be redefined as

$$\sum_{f \in F} \sum_{t \in T_f} \sum_{(i,j) \in E} Y_{ij}^{tf}$$

and the delay function as

$$\sum_{f \in F} \sum_{t \in T_f} \sum_{(i,j) \in E} d_{ij}.Y_{ij}^{tf}$$

To solve the problem with this method, we must rewrite the functions:

$$\min Z = r_1.\sum_{f \in F} \sum_{t \in T_f} \sum_{(i,j) \in E} Y_{ij}^{tf} + r_2.\sum_{f \in F} \sum_{t \in T_f} \sum_{(i,j) \in E} d_{ij}.Y_{ij}^{tf}$$
$$+ r_3.\sum_{f \in F} \sum_{(i,j) \in E} bw_f.\max\left(X_{ij}^{tf}\right)_{t \in T_f} + r_4.\alpha \tag{17}$$

subject to

$$\alpha = \max\left\{\alpha_{ij}\right\}, \text{ where } \alpha_{ij} = \frac{\sum_{f \in F} bw_f.\max\left(X_{ij}^{tf}\right)_{t \in T_f}}{c_{ij}}$$

$$\sum_{(i,j) \in E} X_{ij}^{tf} = 1, t \in T_f, f \in F, i = s$$

$$\sum_{(j,i) \in E} X_{ji}^{tf} = -1, i = t, t \in T_f, f \in F$$

$$\sum_{(i,j) \in E} X_{ij}^{tf} - \sum_{(j,i) \in E} X_{ji}^{tf} = 0, t \in T_f, f \in F, i \neq s, i \neq t$$

$$\sum_{f \in F} bw_f . \max\left(X_{ij}^{tf}\right)_{t \in T_f} \le c_{ij}, (i,j) \in E$$

$$Y_{ij}^{tf} = \left\lceil X_{ij}^{tf} \right\rceil = \begin{cases} 0, X_{ij}^{tf} = 0 \\ 1, 0 < X_{ij}^{tf} \le 1 \end{cases}$$

$$X_{ij}^{tf} = \left\{0,1\right\}, 0 \le r_i \le 1, \sum_{i=1}^{4} r_i = 1$$

Example

We will transmit a single flow f_1 with a transmission rate of 2048 Kbps from origin node 0 to destination nodes 8 and 12.

Table 4.23 shows one solution obtained.

With this function we can find feasible and optimal solutions for the problem when it is necessary to divide the flow into different subflows.

As in the unicast transmission, this mathematical model can be better expressed in multicast transmission using the subflow

Table 4.23   Optimal Solutions in Multicast

Sol	r_1, r_2, r_3, r_4	f_1 (hops)	f_2 (ms)	f_4 (Kbps)	f_7 (%)	Flow Fraction Tree
1	$r_1 = 0.1$ $r_2 = 0.1$ $r_3 = 0.1$ $r_4 = 0.7$	14	166	9728	100	f_1 75% {(0, 2), (2, 7), (7, 8), (7, 13), (13, 12)} f_1 25% {(0, 3), (3, 10), (10, 12), (12, 13)}

concept. In this case, the set of multicast flows is denoted by F. For every flow $f \varepsilon$, F can be divided into K_f subflows and, once normalized, can be denoted as f_k, where $k = 1, ..., |K_f|$. Under this denomination, f_k shows the fraction of the multicast flow $f \varepsilon F$ that is carried by subflow k. For every flow $f \varepsilon F$ we have an origin node $s_f \varepsilon N$ and a set of destination nodes $T_f \subset N$ associated to such flow. In this case, a destination node t of a flow meets with $t \varepsilon T_f$ and also $T = \bigcup_{f \in F} T_f$.

One must include the constraint $\sum_{k=1}^{|k_f|} f_k = 1$ to normalize the values f_k.

In this mathematical formula the vector of variables X_{ij}^{tf} is replaced by the new denotation $X_{ij}^{tf_k}$. $X_{ij}^{tf_k}$ denotes the fraction of the flow f transmitted by subflow k with destination node t through link (i, j). Similarly, it is redefined $Y_{ij}^{tf_k}$, where

$$Y_{ij}^{tf_k} = \left[X_{ij}^{tf_k} \right] = \begin{cases} 0, X_{ij}^{tf_k} = 0 \\ 1, 0 < X_{ij}^{tf_k} \leq 1 \end{cases}$$

For the case of multicast flow transmission through different subflows it is necessary to add a new constraint, a subflow uniformity constraint, to ensure that a subflow f_k always carries the same information:

$$X_{ij}^{f_k t} = \begin{cases} X^{f_k} = \max_{t \in T_f, (i,j) \in E} \left\{ X_{ij}^{f_k t} \right\}, & \text{if } Y_{ij}^{f_k t} = 1 \\ 0, & \text{if } Y_{ij}^{f_k t} = 0 \end{cases}, (i, j) \in E, t \in T_f$$

Without this constraint, $X_{ij}^{f_k t} > 0$ may differ from $X_{ij}^{f_k t'} > 0$, and therefore, the same subflow f_k may not carry the same data to different destinations t and t'. As a consequence of this new constraint, mapping subflows to LSPs is easy.

To solve the problem with this method, we must rewrite the functions with the addition of the new index k:

$$\min Z = r_1 \cdot \sum_{f \in F} \sum_{k \in K_f} \sum_{t \in T_f} \sum_{(i,j) \in E} Y_{ij}^{tf_k} + r_2 \cdot \sum_{f \in F} \sum_{k \in K_f} \sum_{t \in T_f} \sum_{(i,j) \in E} d_{ij} \cdot Y_{ij}^{tf_k}$$

$$+ r_3 \cdot \sum_{f \in F} \sum_{k \in K_f} \sum_{(i,j) \in E} bw_f \cdot \max\left(X_{ij}^{tf_k}\right)_{t \in T_f} + r_4 \cdot \alpha \qquad (18)$$

subject to

$$\alpha = \max\left\{\alpha_{ij}\right\}, \text{ where } \alpha_{ij} = \frac{\displaystyle\sum_{f \in F} \sum_{k \in K_f} bw_f \cdot \max\left(X_{ij}^{tf_k}\right)_{t \in T_f}}{c_{ij}}$$

$$\sum_{(i,j) \in E} \sum_{k \in K_f} X_{ij}^{tf_k} = 1, t \in T_f, f \in F, i = s_f$$

$$\sum_{(j,i) \in E} \sum_{k \in K_f} X_{ji}^{tf_k} = -1, i = t, t \in T_f, f \in F$$

$$\sum_{(i,j) \in E} X_{ij}^{tf_k} - \sum_{(j,i) \in E} X_{ji}^{tf_k} = 0, t \in T_f, f \in F, k \in K_f, i_f \neq s_f, i_f \neq t$$

$$X_{ij}^{f_k t} = \begin{cases} X^{f_k} = \max\limits_{t \in T_f, (i,j) \in E}\left\{X_{ij}^{f_k t}\right\}, & \text{if } Y_{ij}^{f_k t} = 1 \\ 0, & \text{if } Y_{ij}^{f_k t} = 0 \end{cases}, (i,j) \in E, t \in T_f$$

$$\sum_{f \in F} \sum_{k \in K_f} bw_f \cdot \max\left(X_{ij}^{tf_k}\right)_{t \in T_f} \leq c_{ij}, (i,j) \in E$$

$$Y_{ij}^{tf_k} = \left\lceil X_{ij}^{tf_k}\right\rceil = \begin{cases} 0, X_{ij}^{tf_k} = 0 \\ 1, 0 < X_{ij}^{tf_k} \leq 1 \end{cases}$$

$$X_{ij}^{tf_k} = \left\{0,1\right\}, 0 \leq r_i \leq 1, \sum_{i=1}^{4} r_i = 1$$

4.6 Multi-Objective Optimization Modeling

In this section we will present the model by means of a real multi-objective scheme.

In this case, functions are handled as independent functions and no approximation is made through a single-objective method. In other words, we are going to rewrite the vector of objective functions instead of a single-objective approximation. Constraints remain exactly the same.

4.6.1 Unicast Transmission

If we are going to optimize the functions hop and delay, the vector of the objective functions would be given by

$$
F\left(X_{ij}^{f}\right) = \left[\sum_{f \in F} \sum_{(i,j) \in E} X_{ij}^{f}, \sum_{f \in F} \sum_{(i,j) \in E} d_{ij}.X_{ij}^{f} \right]
$$

If, in addition, we want to optimize the bandwidth consumption function, the vector of objective functions would be given by

$$
F\left(X_{ij}^{f}\right) = \left[\sum_{f \in F} \sum_{(i,j) \in E} X_{ij}^{f}, \sum_{f \in F} \sum_{(i,j) \in E} d_{ij}.X_{ij}^{f}, \sum_{f \in F} \sum_{(i,j) \in E} bw_{f}.X_{ij}^{f} \right] \quad (19)
$$

subject to

$$
\sum_{(i,j) \in E} X_{ij}^{f} = 1, f \in F, i = s
$$

$$
\sum_{(j,i) \in E} X_{ji}^{f} = -1, i = t, f \in F
$$

$$
\sum_{(i,j) \in E} X_{ij}^{f} - \sum_{(j,i) \in E} X_{ji}^{f} = 0, f \in F, i \neq s, i \neq t
$$

$$\sum_{f \in F} bw_f . X_{ij}^f \le c_{ij}, (i,j) \in E$$

$$X_{ij}^f = \left\{0,1\right\}$$

This way, we can add the functions that we need to optimize without interfering in the analytical model to be solved.

If, in addition, we want to optimize the maximum link utilization, the model would be the following (in this case, it has been necessary to add some constraints):

$$F\left(X_{ij}^{f_k}\right) = \left[\sum_{f \in F}\sum_{k \in K_f}\sum_{(i,j) \in E} Y_{ij}^{f_k}, \sum_{f \in F}\sum_{k \in K_f}\sum_{(i,j) \in E} d_{ij}.Y_{ij}^{f_k}, \sum_{f \in F}\sum_{k \in K_f}\sum_{(i,j) \in E} bw_f.X_{ij}^{f_k}, \alpha\right]$$

(20)

subject to

$$\alpha = \max\left\{\alpha_{ij}\right\}, \text{ where } \alpha_{ij} = \frac{\displaystyle\sum_{f \in F}\sum_{k \in K_f} bw_f.X_{ij}^{f_k}}{c_{ij}}$$

$$\sum_{(i,j) \in E}\sum_{k \in K_f} X_{ij}^{f_k} = 1, f \in F, i = s_f$$

$$\sum_{(j,i) \in E}\sum_{k \in K_f} X_{ji}^{f_k} = -1, i = t_f, f \in F$$

$$\sum_{(i,j) \in E} X_{ij}^{f_k} - \sum_{(j,i) \in E} X_{ji}^{f_k} = 0, f \in F, k \in K_f, i \ne s_f, i \ne t_f$$

$$\sum_{f \in F}\sum_{k \in K_f} bw_f.X_{ij}^{f_k} \le c_{ij}, (i,j) \in E$$

$$Y_{ij}^{f_k} = \left\lceil X_{ij}^{f_k} \right\rceil = \begin{cases} 0, X_{ij}^{f_k} = 0 \\ 1, 0 < X_{ij}^{f_k} \le 1 \end{cases}$$

$$X_{ij}^{f_k} = \{0,1\}, 0 \le r_i \le 1, \sum_{i=1}^{4} r_i = 1$$

From these models we can include any other objective function that we wish to optimize or any other constraint that we need to add.

4.6.2 Multicast Transmission

If we are going to optimize the functions hop and delay, the vector of the objective functions would be given by

$$F\left(X_{ij}^{tf}\right) = \left[\sum_{f \in F} \sum_{t \in T_f} \sum_{(i,j) \in E} X_{ij}^{tf} , \sum_{f \in F} \sum_{t \in T_f} \sum_{(i,j) \in E} d_{ij}.X_{ij}^{tf} \right] \qquad (21)$$

If, in addition, we want to optimize the bandwidth consumption function, the vector of the objective functions would be given by

$$F\left(X_{ij}^{tf}\right) = \left[\sum_{f \in F} \sum_{t \in T_f} \sum_{(i,j) \in E} X_{ij}^{tf} , \sum_{f \in F} \sum_{t \in T_f} \sum_{(i,j) \in E} d_{ij}.X_{ij}^{tf}, \sum_{f \in F} \sum_{(i,j) \in E} bw_f.\max\left(X_{ij}^{tf}\right)_{t \in T_f} \right]$$

$$(22)$$

subject to

$$\sum_{(i,j) \in E} X_{ij}^{tf} = 1, t \in T_f, f \in F, i = s$$

$$\sum_{(j,i) \in E} X_{ji}^{tf} = -1, i = t, t \in T_f, f \in F$$

$$\sum_{(i,j)\in E} X_{ij}^{tf} - \sum_{(j,i)\in E} X_{ji}^{tf} = 0, t \in T_f, f \in F, i \neq s, i \neq t$$

$$\sum_{f\in F} bw_f . \max\left(X_{ij}^{tf}\right)_{t\in T_f} \leq c_{ij}, (i,j) \in E$$

$$X_{ij}^{tf} = \left\{0,1\right\}$$

And if we want to optimize the maximum link utilization function, the model would be the following:

$$F\left(X_{ij}^{tf_k}\right) = \left[\sum_{f\in F}\sum_{k\in K_f}\sum_{t\in T_f}\sum_{(i,j)\in E} Y_{ij}^{tf_k}, \sum_{f\in F}\sum_{k\in K_f}\sum_{t\in T_f}\sum_{(i,j)\in E} d_{ij}.Y_{ij}^{tf_k}, \right.$$

$$\left. \sum_{f\in F}\sum_{k\in K_f}\sum_{(i,j)\in E} bw_f . \max\left(X_{ij}^{tf_k}\right)_{t\in T_f}, \alpha \right] \tag{23}$$

subject to

$$\alpha = \max\left\{\alpha_{ij}\right\}, \text{where } \alpha_{ij} = \frac{\displaystyle\sum_{f\in F}\sum_{k\in K_f} bw_f . \max\left(X_{ij}^{tf_k}\right)_{t\in T_f}}{c_{ij}}$$

$$\sum_{(i,j)\in E}\sum_{k\in K_f} X_{ij}^{tf_k} = 1, t \in T_f, f \in F, i = s_f$$

$$\sum_{(j,i)\in E}\sum_{k\in K_f} X_{ji}^{tf_k} = -1, i = t, t \in T_f, f \in F$$

$$\sum_{(i,j)\in E} X_{ij}^{tf_k} - \sum_{(j,i)\in E} X_{ji}^{tf_k} = 0, t \in T_f, f \in F, k \in K_f, i_f \neq s_f, i_f \neq t$$

$$X_{ij}^{fkt} = \begin{cases} X^{f_k} = \max\limits_{t \in T_f,(i,j) \in E} \left\{ X_{ij}^{fkt} \right\}, & \text{if } Y_{ij}^{fkt} = 1 \\ 0, & \text{if } Y_{ij}^{fkt} = 0 \end{cases}, (i,j) \in E, t \in T_f$$

$$\sum_{f \in F} \sum_{k \in K_f} bw_f \cdot \max\left(X_{ij}^{tf_k} \right)_{t \in T_f} \le c_{ij} \quad ,(i,j) \in E$$

$$Y_{ij}^{tf_k} = \left\lceil X_{ij}^{tf_k} \right\rceil = \begin{cases} 0, X_{ij}^{tf_k} = 0 \\ 1, 0 < X_{ij}^{tf_k} \le 11 \end{cases}$$

$$X_{ij}^{tf_k} = \{0,1\}, 0 \le r_i \le 1, \sum_{i=1}^{4} r_i = 1$$

4.7 Obtaining a Solution Using Metaheuristics

To solve multi-objective optimization problems applied to computer networks, one can use certain types of metaheuristics. In this section we will discuss the source codes developed in C language of a metaheuristic, which solve the optimization problem under a simple solution and do not have the intention of being the most effective and efficient algorithm to provide this solution. If readers want to analyze advanced algorithms to solve these types of problems, they can review some of the works and books recommended in the bibliography. Even though some of the metaheuristics mentioned in Chapter 2 can be used, we will concentrate on working with the Multi-objective Evolutionary Algorithm (MOEA). Specifically, we will work with the methodology proposed by the Strength Pareto Evolutionary Algorithm (SPEA) to solve the stated problems.

In this section we will obtain a computational solution to the problem previously stated for unicast as well as multicast transmissions.

To execute the evolutionary algorithm, one must first define how the solutions are going to be represented — in other words, what is the structure of a chromosome. In addition, one must define how the search process of the initial population is going to take place, how the chromosome selection process will be developed, what the crossover function will be, and, lastly, what the mutation function will be.

4.7.1 Unicast for the Hop Count and Delay Functions

Figure 4.30 shows the topology that we will use in the unicast case.

4.7.1.1 Coding of a Chromosome

Figure 4.31 shows a chromosome (or a solution) corresponding to a path from origin node s to destination node t. It is a vector that contains all the nodes that form that path. Overall, the path represented by Figure 4.31 would be given by nodes s, n_1, n_2, ..., n_i, ..., t.

Figure 4.32 shows a representation of chromosomes corresponding to the five solutions that can be considered in Figure 4.30.

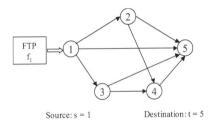

Source: s = 1   Destination: t = 5

Figure 4.30 Unicast topology.

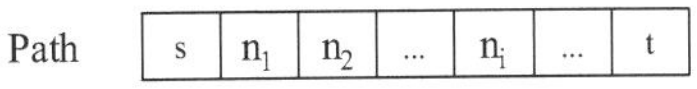

Path | s | n_1 | n_2 | ... | n_i | ... | t |

Figure 4.31 Representation of a chromosome.

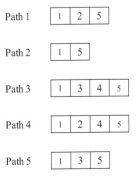

Figure 4.32 Solutions.

4.7.1.2 Initial Population

To establish the initial population we use the breadth first search (BFS) graph search algorithm, which was explained in Chapter 1. Here, the algorithm will be executed probabilistically. Given a path to destination t, if probability P is not met, such path will not be explored. Because it is probabilistic, the search of this algorithm would not be exhaustive. The idea is to find several paths from origin node s to destination node t to consider them as the initial population. The paths not found by algorithm BFS will be found by the crossover and mutation procedures of the MOEA. This way, we will obtain a population of feasible paths to solve the problem.

4.7.1.3 Selection

Selection of a chromosome is done probabilistically. The value associated with the selection probability is given by the chromosome's fitness, which tells how good this chromosome is compared with others in its population. This algorithm is simple; the idea is to select all chromosomes that are not dominated in a generation, that is, in an iteration of the algorithm. These chromosomes form an elitist population called P_{nd}. The other chromosomes are dominated by population P_{nd} and form a population called P. The fitness value of each chromosome, both in population P_{nd} as in P, is calculated by the SPEA to ensure that the worst fitness value of the elitist P_{nd} is better than the best value in population P. Hence, chromosomes in P_{nd} have a greater probability of being selected. Once fitness values are found for all chromosomes, one uses the roulette selection operator or the binary tournament (explained in Chapter 2) on the junction of both populations P_{nd} and P to select one chromosome at a time every time needed.

4.7.1.4 Crossover

The crossover and mutation functions are subject to combinatorial analysis to find new chromosomes from existing chromosomes.

We propose the following scheme as the crossover operator. Using the selection procedure, one selects two chromosomes from which the offspring chromosomes will be produced. Next, one checks the crossover probability to determine whether the crossover will take place. If possible, one selects a cutting point, which can be obtained randomly, in any position past the origin node s and prior to destination node t. The crossover function considered in this book only has one cutting point, but one could consider crossover functions with more than one cutting point.

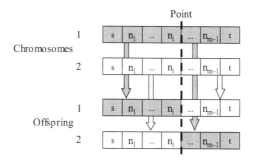

Figure 4.33   Crossover operator.

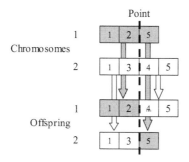

Figure 4.34   Crossover example.

Once the cutting point is determined, one obtains the offspring in the following way: the left part of the first parent chromosome is selected, the right part of the second parent chromosome is selected, and then both are combined to produce a new offspring. An additional crossover can be done to produce another offspring; in this case, the left part of the second parent chromosome is combined with the right side of the first parent chromosome (see Figure 4.33). Lastly, one must verify that the offspring created (one or two) are really feasible solutions to the problem.

Figure 4.34 shows how from two existing paths in the initial population (parent chromosomes) one obtains two new paths (offspring chromosomes). The crossover point has been selected in the second position. Finally, we must verify that both paths are feasible in the topology shown in Figure 4.10.

4.7.1.5 Mutation

We propose the following scheme as the mutation operator. By means of the selection method we obtain a chromosome. Next, one checks the

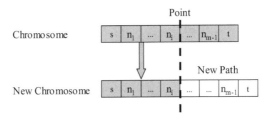

Figure 4.35 Mutation operator.

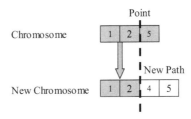

Figure 4.36 Mutation example.

probability of mutation to verify whether the mutation will take place. If positive, one selects a cutting point to perform it. This point can be deterministic, for example, always in the center of the chromosome, or it can be selected probabilistically, this being most recommended. The part of the chromosome between origin node s and cutting point n_i is maintained, and as from node n_i, one performs, for example, through the BFS algorithm, the search of a new path between node n_i and the destination node t.

Figure 4.35 shows how the mutation operator works.

Figure 4.36 shows how in the topology analyzed (Figure 4.30), from existing path {(1, 2), (2, 5)}, one obtains the new path {(1, 2), (2, 4), (4, 5)}. In this case, the mutation point is in the second position, and through, for example, the BFS algorithm, one has found the new path {(2, 4), (4, 5)} between nodes 2 and 5. Finally, we must verify that this path corresponds to a feasible path in the topology considered.

4.7.2 Multicast for the Hop Count and Delay Functions

Figure 4.37 shows the topology that we will use in the multicast case.

4.7.2.1 Coding of a Chromosome

Figure 4.38 shows a chromosome (or a solution) corresponding to a tree from origin node s to each of the destination nodes t. It is a vector with

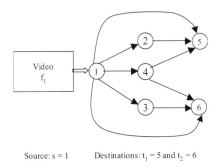

Source: s = 1   Destinations: $t_1 = 5$ and $t_2 = 6$

Figure 4.37   Multicast topology.

Tree | P_1 | # | P_2 | ... | P_i | ... | # | P_n |

Figure 4.38   Representation of a chromosome.

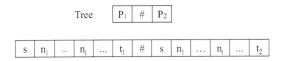

Figure 4.39   Representation of the nodes of each path.

n paths (one to each of the destination nodes). Each of these n paths P_i, with $i = 1, ..., n$ is separated by the special character #.

As we have considered for the unicast case, every path P_i is formed by the nodes that are part of such path. Figure 4.39 shows the tree that corresponds to a multicast transmission with two destination nodes (t_1 and t_2).

The tree could also be represented using matrices. In this case, each row represents a path with one destination node t.

There are other ways to represent these trees, for example, with dynamic structures that would optimize the use of memory in PCs. In this book we will use the tree representation shown in Figure 4.39 and Figure 4.40.

Figure 4.41 shows some of the trees that can be obtained in the topology of Figure 4.37 for a multicast flow from origin node 1 to destination nodes 5 and 6.

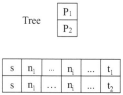

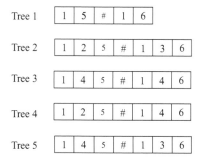

Figure 4.40   Representation in matrix form.

Tree 1 | 1 | 5 | # | 1 | 6 |

Tree 2 | 1 | 2 | 5 | # | 1 | 3 | 6 |

Tree 3 | 1 | 4 | 5 | # | 1 | 4 | 6 |

Tree 4 | 1 | 2 | 5 | # | 1 | 4 | 6 |

Tree 5 | 1 | 4 | 5 | # | 1 | 3 | 6 |

Figure 4.41   Solutions.

4.7.2.2 Initial Population

To establish the initial population, we use the BFS algorithm probabilistically. Here, the idea is to find the initial population formed by several trees from origin node s to the set of destination nodes t.

4.7.2.3 Selection

The chromosome selection process (which in the multicast case represents a tree) can be done the same way as for the unicast case.

4.7.2.4 Crossover

By means of the crossover and mutation functions one can perform combinatorial analysis to find new chromosomes from existing chromosomes.

As the crossover operator we propose the following scheme. Using the selection procedures, select two chromosomes with which one will create the offspring chromosome. Next, check the crossover probability to verify whether the crossover will effectively take place. In the multicast

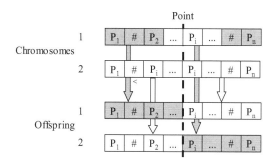

Figure 4.42 Crossover operator.

case, one can define two crossover functions. The first one consists of randomly selecting the path that we will use as the cutting point. As previously mentioned, we are using a single cutting point, but more than one could be used.

The offspring are created in the following way (Figure 4.42): from the first parent chromosome select the left part of the chromosome, and from the second parent chromosome select the right part; then combine the two parts to create a new offspring. The crossover taking place here is interchanging paths of the two trees and also ensuring that both offspring trees have paths to reach all destination nodes t. An offspring could also be obtained by combining the left part of the second parent chromosome with the right part of the first parent chromosome.

The second crossover function consists of selecting a path i of each chromosome, with the constraint that all of these paths have the same node t as destination. In this case, the crossover is done like in the unicast case, dividing each of the chromosome paths by a randomly selected cutting point and interchanging both parts of the path between the two partial chromosomes considered.

In Figure 4.43 one can see how the two paths i (1 and 2 in Figure 4.43) of each of the two chromosomes selected crossed, as from the selected cutting point.

Figure 4.44 shows how, for the multicast case, from the two existing trees in the initial population (parent chromosomes) one obtains two new trees (offspring chromosomes) by means of the first crossover function. We consider parent chromosomes trees [{(1, 2), (2, 5)}, {(1, 3), (3, 6)}] and [{(1, 4), (4, 5)}, {(1, 4), (4, 6)}], and the crossover point is selected after the first path. Doing the crossover function explained previously, the offspring chromosomes would be trees [{(1, 2), (2, 5)}, {(1, 4), (4, 6)}], and [{(1, 4), (4, 5)}, {(1, 3), (3, 6)}].

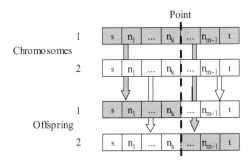

Figure 4.43   Crossover operator 2.

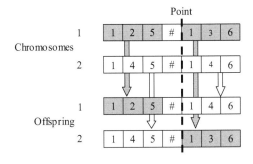

Figure 4.44   Crossover example.

4.7.2.5 Mutation

We propose the following scheme as the mutation operator. Using the selection method, we obtain a chromosome. Next, check the probability of mutation to verify whether the mutation will take place. If positive, one selects a path i of the tree with a destination node t to perform it. In this case, this point can be determined deterministically or probabilistically. The part of the chromosome between origin node s and the cutting point in node n_k remains, and next, from node n_k, the search takes place, for example, through the BFS algorithm, for a new path from node n_k to destination node t_i. As we can see, the mutation function for the multicast case is similar to that for the unicast case. The only difference is that one must first select one of the paths with destination node t.

Figure 4.45 shows how the mutation operation works.

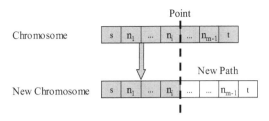

Figure 4.45 Mutation operator.

4.7.3 Unicast Adding the Bandwidth Consumption Function

As mentioned in Section 4.5.3, when the bandwidth consumption function is added, it is possible that one solution may consist of more than one path. For this reason, the chromosome specified in Section 4.7.1 must be modified to represent a solution in which more than one path is found as a solution to the problem. In these solutions, a flow may be carried completely (100 percent) by one path, but another flow from the same origin to the same destination might also be carried completely (100 percent) by a different path, and therefore, one must redefine this chromosome.

4.7.3.1 Coding of a Chromosome

Figure 4.46 shows a chromosome that can represent more than one path. As in the multicast case, the character # separates the different paths, but in this case, each path goes from origin node s to destination node t, transmitting a specific flow. In other words, path P_1 carries flow f_1, path P_i carries flow f_i, and so on. With this type of chromosome we can have a solution in which the same path carries more than one flow.

Now, as can be seen in Figure 4.47, every path P_i consists of the nodes through which the packets from origin node s to destination node t must be transmitted.

4.7.3.2 Initial Population

To establish the initial population, we can once again use the probabilistic BFS algorithm.

Chromosome	P_1	#	P_2	...	P_i	...	#	P_n

Figure 4.46 Representation of a chromosome.

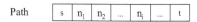

Figure 4.47 Representation of a chromosome path.

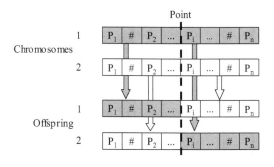

Figure 4.48 Crossover operator 1.

4.7.3.3 Selection

This process is done in the same way as the SPEA, explained previously in Section 2.4.

4.7.3.4 Crossover

As crossover operator we propose two possible crossover functions. The first one consists of selecting two parent chromosomes, randomly selecting a crossing point, and combining each of the parts of the parent chromosomes (Figure 4.48). Another possibility would be to select $|F|$ chromosomes and create offspring selecting a path P_i associated with flow f_i from each parent chromosome. This way, with the $|F|$ parent chromosomes one obtains a new offspring.

The second crossover function consists of selecting two chromosomes and randomly selecting a path as the crossover point (Figure 4.49), using the same crossover operator explained in Section 4.7.1.

4.7.3.5 Mutation

In this case, we will apply the same operator explained in Section 4.7.1 as the mutation operator. One selects a chromosome, randomly selects a position on this chromosome, which corresponds to a path, and applies the mutation operator explained in Section 4.7.1. For example, Figure 4.50 shows that in the chromosome selected one has chosen the position corresponding to path P_2, and subsequently, on this path (P_2) one selects

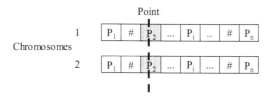

Figure 4.49 Path selection.

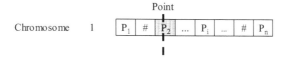

Figure 4.50 Path selection.

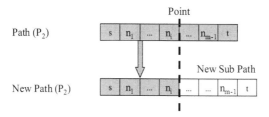

Figure 4.51 Mutation operator.

a point from which one performs the search of a new path to destination node *t* (Figure 4.51).

This way, by means of the mutation operator from an existing path, we would find new paths for a flow *f*.

4.7.4 Multicast Adding the Bandwidth Consumption Function

As was seen when the bandwidth consumption function was added to multicast transmissions, it is possible for a solution to consist of more than one tree. With these types of solutions, a multicast flow can be completely carried (100 percent) by one tree, but another flow could also be carried completely (100 percent) by a different tree. For this reason, this chromosome must be redefined the same way as the unicast case was redefined.

4.7.4.1 Coding of a Chromosome

Figure 4.52 shows a chromosome that can represent more than one tree, and the character # separates the different trees. In this case, for each multicast flow f there is a tree that goes from origin node s to a set of destination nodes T_f. In other words, tree T_1 carries flow f_1, tree T_i carries flow f_i, and so on. In this case, we can have a chromosome in which the same tree carries more than one multicast flow.

Each tree consists of $|T_f|$ paths, as many as the number of destination nodes that flow f has. Figure 4.53 shows a tree T_i that goes from origin node s to a set of destination nodes T_f; it is formed by a vector where each path with a destination node $t \in T_f$ is separated by the special character $*$. Hence, if we have $|T_f|$ destination nodes, we would represent the $|T_f|$ paths toward each of the destination nodes.

In addition, as with the unicast case, every path P_i is formed by the nodes that make up such path. Figure 4.54 shows that every path consists of the set of nodes and, for this specific case, where only two destination nodes exist.

4.7.4.2 Initial Population

To determine the initial population, we will once again use the probabilistic BFS algorithm, and for every flow f, we will obtain a tree.

4.7.4.3 Selection

The selection process of a chromosome in the multicast case can be done in the same way as in the unicast case.

Chromosome | T_1 | # | T_2 | ... | T_i | ... | # | T_n |

Figure 4.52 Representation of a chromosome (trees).

Tree | P_1 | * | P_2 | ... | P_i | ... | * | P_{Tf} |

Figure 4.53 Representation of a chromosome (paths of the tree T_i).

Path 1 | s | n_1 | ... | n_i | ... | t_1 |
Path 2 | s | n_1 | ... | n_i | ... | t_2 |

Figure 4.54 Representation of the paths.

4.7.4.4 Crossover

As crossover operator we propose three possible crossover functions. The first consists of selecting two parent chromosomes, selecting a crossover point, and combining each of the parts (Figure 4.55). With this crossover function we can exchange different trees in the offspring chromosome.

The second crossover function consists of selecting two parent chromosomes and randomly selecting a crossover point (Figure 4.56). In this case, we would be selecting a tree from each chromosome for the same flow. Subsequently, one randomly selects a crossover point within each tree, and we would obtain the paths to the different destinations of the flow. Finally, one combines each of the parts of the parent chromosomes, that is, of the paths of the selected tree (Figure 4.57).

The third crossover function consists of selecting two chromosomes, randomly obtaining a tree, and, from this selected tree, randomly obtaining a path to which we will apply the same crossover operator that was explained in Section 4.7.1.

In Figure 4.58 we see an example where we have selected path P_2 of tree T_2 of multicast flow f, and from this point on we apply the crossover operator explained in Section 4.7.1.

Figure 4.59 shows the third crossover function.

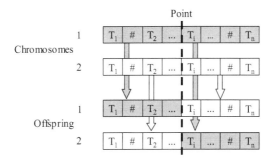

Figure 4.55   Crossover operator 1.

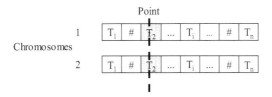

Figure 4.56   Tree selection.

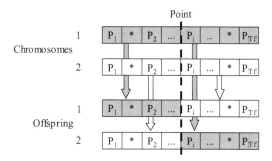

Figure 4.57   Crossover operator 2.

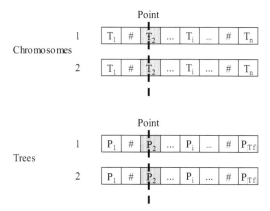

Figure 4.58   Tree and path selection.

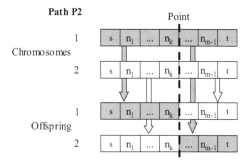

Figure 4.59   Crossover operator 3.

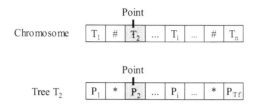

Figure 4.60 Tree and path selection.

4.7.4.5 Mutation

As mutation operator, in this case we will apply the same operator as in Section 4.7.1. We select a chromosome and randomly select a position of this chromosome, tree T_2 in the example in Figure 4.60. Within tree T_2 we randomly select a path (for example, P_2) and, subsequently, apply the mutation operator explained in Section 4.7.1. This way, by means of the mutation operator, we would find new paths for a flow f from an existing path.

4.7.5 Unicast Adding the Maximum Link Utilization Function

By adding the maximum link utilization function, it is possible that a single flow f may have more than one path from origin node s to destination node t, and that every path carries a fraction of flow f. For this reason, one must redefine the chromosome in order for it to represent these fractions of flow f.

4.7.5.1 Coding of a Chromosome

Figure 4.61 shows a chromosome that can represent more than one path for every flow f. The main structure of the chromosome consists of a vector for the different unicast flows that are transmitted. In addition, every flow f consists of subflows, and every subflow consists of a path and the percentage of flow f that it carries. As in the previous cases, every path consists of the set of nodes from origin node s to destination node t. With this chromosome a flow can be carried through several paths, and every path carries a fraction of this flow.

Every solution (or chromosome) consists of $|F|$ flows, every flow f consists of $|K_f|$ subflows, and every subflow f consists of a path from origin node s to destination node t and of the fraction of flow f_k that it is carrying of flow f (Figure 4.61).

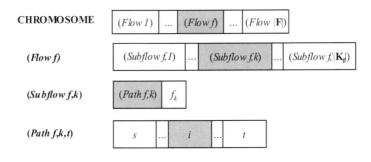

Figure 4.61   Representation of a chromosome.

4.7.5.2 Initial Population

In this case, we can use the same BFS algorithm, but now seeking paths for subflows of f. For every flow f one can randomly select the number of subflows that will be created initially to carry 100 percent of the flow, and subsequently, for every subflow with algorithm BFS we find a path from origin node s to destination node t and the percentage of flow that this path will carry.

4.7.5.3 Selection

This process is done in the same way as the SPEA, previously explained.

4.7.5.4 Crossover

Two crossover functions can be used as the crossover operator. The first consists of selecting $|F|$ parent chromosomes, sequentially selecting a flow from every parent chromosome, and combining all $|F|$ parent chromosomes to create an offspring chromosome. With this process one can obtain several offspring, and we take different flows from every parent chromosome (Figure 4.62).

The second crossover function consists of selecting two chromosomes and the flow f that will be used for the crossover function. Subsequently, one takes the same flow f of both chromosomes, randomly selects two crossover points, and crosses them as shown in Figure 4.63. Here, two offspring have been created. By means of this crossover function we would be changing the paths for every flow f, and additionally, we would be reducing or increasing the number of paths or subflows through which flow f will be transmitted. Lastly, we must normalize the values of the

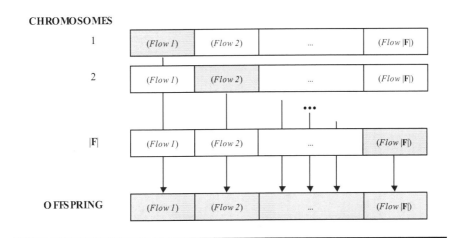

Figure 4.62 Crossover operator 1.

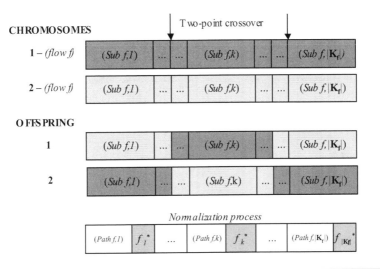

Figure 4.63 Crossover operator 2.

fraction to transmit for every subflow in order for the sum of the fraction to continue to be 1.

4.7.5.5 Mutation

In this case, we will apply as mutation operator the same operator that we explained in Section 4.7.1. We select a chromosome, then randomly select a position of this chromosome, which corresponds to the path used

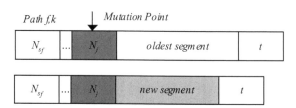

Figure 4.64 Mutation operator 1.

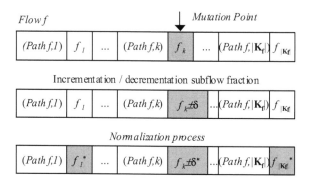

Figure 4.65 Mutation operator 2.

by a subflow, and, subsequently, apply the mutation operator explained in Section 4.7.1, which through a randomly selected point seeks a new path from that node to destination node t. This mutation operator is shown in Figure 4.64.

Through the second function of the mutation operator what one seeks is to be able to change the value of the fraction of flow to be transmitted by that path (Figure 4.65). It can happen that with this function a path (or subflow) will disappear when the value of the fraction is zero. With this operator, we select a chromosome, randomly select a flow f, and also randomly select a subflow. To the value of the fraction of this subflow f_k one will randomly add a value δ. Lastly, one must normalize the values of the fractions transmitted by each subflow for the sum of these fractions to be 1.

4.7.6 Multicast Adding the Maximum Link Utilization Function

If we add the maximum link utilization function, it may be possible that only one multicast flow f can have more than one tree from origin node s to a set of destination nodes t, and that every tree carries a fraction of

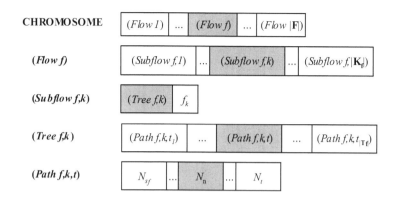

Figure 4.66 Representation of a chromosome.

flow *f*. This chromosome must be able to demonstrate that every multicast flow may have more than one tree to carry fractions of flow *f*.

4.7.6.1 Coding of a Chromosome

In Figure 4.66 one can see a chromosome that represents more than one tree for every multicast flow *f*. The chromosome consists of one vector for the different multicast flows that will be transmitted. Every flow *f* in turn consists of subflows. Every subflow consists of a tree and of the percentage of flow *f* it will carry. Every tree consists of the paths toward every one of the destination codes. And lastly, every path consists of a set of nodes in its route from origin node *s* to destination node *t*. With this type of chromosome we can have as a solution a flow that can be carried through several trees, and that every tree carries a fraction of such flow.

As can be seen in Figure 4.66, every solution (or chromosome) consists of $|F|$ flows, every flow *f* consists of $|K_f|$ subflows, and every subflow of *f* consists of a tree from origin node *s* to every one of the destination nodes T_f and of the fraction of flow f_k that it is carrying of flow *f*. Lastly, every path consists of the set of nodes that are part of the route from origin node *s* to destination node *t*.

4.7.6.2 Initial Population

In this case, we can use the same probabilistic BFS, but this time seeking trees for the subflows of *f*. For every flow *f* carried, one can randomly select the number of subflows that will be created initially to carry 100 percent of the flow. With algorithm BFS we can find for all subflows the paths from origin node *s* to every one of the destination nodes *t* and the

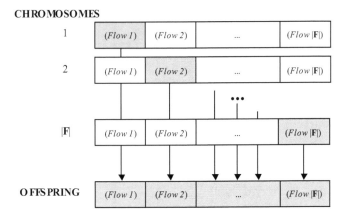

Figure 4.67 Crossover operator 1.

percentage of flow that this tree will carry. The sum of all subflows for a multicast flow *f* must total 1.

4.7.6.3 Selection

This process is done in the same way as the SPEA, which was previously explained.

4.7.6.4 Crossover

We propose two possible crossover functions as the crossover operator. The first one consists of selecting parent $|F|$ chromosomes, sequentially selecting one flow of every parent chromosome, and combining the $|F|$ parent chromosomes to create an offspring chromosome. With this process one can obtain several offspring, taking different flows from every parent chromosome (Figure 4.67).

The second crossover function consists of selecting two chromosomes and one flow *f* that will be used for the crossover function. Subsequently, one takes the same flow *f* of both chromosomes and randomly selects two crossover points to later cross them as shown in Figure 4.68. In this case, one has created two offspring. Through this crossover function we would be changing the paths for every flow *f* and also would be reducing or increasing the number of paths or subflows through which flow *f* will be transmitted. Lastly, we must normalize the values of the fraction to be transmitted by every subflow in order for the sum of the fractions to be 1.

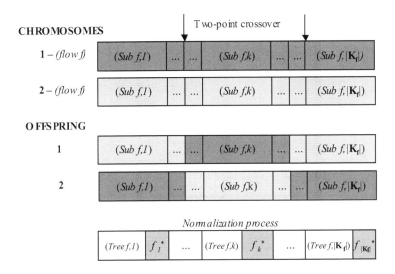

Figure 4.68   Crossover operator 2.

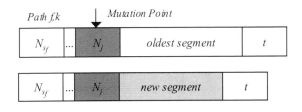

Figure 4.69   Mutation operator 1.

4.7.6.5 Mutation

As the mutation operator in this case, we will apply the same operator as in Section 4.7.1. We select a chromosome, randomly select a position of this chromosome, which corresponds to a flow f, randomly select a subflow, which corresponds to a tree, and, lastly, randomly select one or several paths. To each of these paths one subsequently applies the mutation operator explained in Section 4.7.1, with which, once a node is randomly selected, one seeks a new path from that node to destination node t. This mutation operator is shown in Figure 4.69.

Through the second function of the mutation operator what one seeks is to change the value of the fraction of flow that will be transmitted by that path (Figure 4.70). With this function it can happen that a tree or subflow disappears if the value of the fraction is zero. For this operator, we select a chromosome, then we randomly select a flow f and also

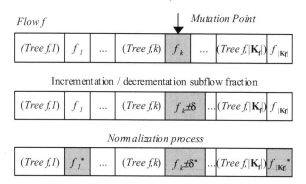

Figure 4.70 Mutation operator 2.

randomly select a subflow. To the value of the fraction of this subflow f_k one will randomly add or subtract a value δ. Finally, one must normalize the values of the fractions transmitted by every subflow.

To conclude, we want to mention that three different chromosomes have been defined with their respective evolutionary operators. Every type of chromosome solves the optimization of different objective functions. Other objective functions mentioned in this book (or not mentioned) can be added to the chromosomes that we have analyzed, and even with slight changes, the source codes may serve to execute the respective optimization processes both at the traditional methods level and at the metaheuristic by means of the MOEA.

Chapter 5

Multi-Objective Optimization in Optical Networks

5.1 Concepts

Even though optical fiber was manufactured during the 1970s, the complete capacity of this technology is not being used, largely due to the fact that in the past there were no applications that demanded facilities for large traffic. Lately, progress in the computer and network world, such as videoconferences, telemedicine, video, and image and text files distribution, demands the development of optical networks in extensive areas.

Not all the networks that use optical fiber are optical networks. We will define optical networks as telecommunications networks whose transmission links are optical fibers and whose architecture is designed to make the most of fiber characteristics. The objective is to have a network with a totally optical core (optical network node (ONN)), with the links connected through optical nodes, and in the end of such core there should be access stations to the network, called network attached storage (NAS), which serve as the interface, so that the user terminals and other optical systems can connect to the optical network.

NAS provides terminal points (emitters and receptors) for travel of the optical signals within the physical layer. Communications outside of the purely optical part of the network continue in an electrical manner, whether ending in a final accessory (for example, user terminals) or going through electronic commutation equipment (for example, Asynchronous Transfer Mode (ATM) switches). The ONNs carry out commutation and routing functions, which control the travel of the optical signals, and they are configured to create emitting/destination connections.

The performance of a network is constrained by the quantity and functionality of its physical resources. Below, we will explain the different functions carried out in an optical network, emphasizing the role of optical resources for supplying connectivity and flow of information.

In terms of hardware, the interface between the logical layer and the physical layer is found in the external ports (electrical) and NAS. As indicated, the optical connections end in the access stations and serve as the interface between the electronic and optical equipment. On the optical side, each NAS is connected to an ONN through an access link that consists of one or more pairs of fiber. The optical signals are interchanged between the station and the network through the station transmitters' receptors. Each node is connected to its neighboring nodes through a pair of fibers, which constitute the intermodal links of the network. A very long link can have one or more optical amplifiers to compensate the attenuation experienced by the fiber.

The sublayers of the optical layer work according to the way in which the optical spectrum is partitioned. The smallest entity is the partition in channels, each one of which is assigned a different wavelength. These are the basic information bearers in the physical layer. It is expected that λ channels will be independently commuted by the ONNs.

Each point-to-point optical connection is carried out through a λ created on two steps. The first step in a channel assigns the λ wavelength to the transmitter of the emitting node and to the receptor of the destination node. The second step establishes an optical route (OP) through a node sequence to carry this wavelength from the source to the destination.

The unidirectional connection between external nodes of a pair of NASs is called logic connection (LC). This connection carries a logical signal in a determined electrical format. All the LCs use the resources of the physical layer, and each LC is transmitted through an optical connection by the transmission channel. The transmission channel carries out the conversion of the logical signal into a transmission signal. This conversion is carried out at the transmission processor (TP). The same inverse option is carried out in the reception processor (RP). Additionally, the transmission signal must be limited to the bandwidth that can be managed by the optical transmitter (OT) and the optical receptor (OR), and the spectrum allowed by the channel that carries the signal. The main reason for the optical layer to divide into sublayers is the operation of the multiplexer — in other words, the handling of multiple accesses and the commuting.

5.1.1 Multiplexing of the Network

A large number of simultaneous connections can be supported in a link of an optical network, using the multiplexing technique. There are different multiplexing techniques. Among them we can mention multiplexing by division of time, space, and frequency. In this chapter we will suppose that we use dense wavelength division multiplexing (DWDM). DWDM uses a smaller separation between the wavelengths than WDM.

Using DWDM, each fiber will have different connections in different wavelengths (channels). The wavelengths have to be sufficiently separated to avoid them from superimposing themselves, which would produce interference in the optical receiver signal. That is why in DWDM the wavelengths are normally separated among themselves by 1 or 0.1 nm.

A network that uses DWDM with channels in m wavelengths, λ_1, λ_2, ..., λ_m, can be considered at the level of point-to-point connection as m copies in the network, each one with the same physical topology. An optical connection between two stations uses an optical route in the network copy corresponding to channel λ assigned to that connection. In an optical network, a route that communicates emitter/destination must use the same channel λ for all the links in the route.

5.1.2 Multiprotocol λ Switching Architecture (MPλS)

MPλS architecture was designed for optical networks by the Internet Engineering Task Force (IETF) based on Multi-Protocol Label Switching (MPLS). MPS tries to correlate an MPLS label to a specific wavelength in such a way that the optical nodes do not need to process the MPLS label, but can make decisions based on the wavelengths. This allows switching without the need for converting the optical signal to an electric signal, because there is no need to read contents of the package. A generalized MPLS (GMPLS) is a set of technical specifications of how MPLS would work under different technologies in the network, for example, the optical networks.

5.1.3 Optical Fiber

As mentioned previously, not all networks that use optical fiber are optical networks, but all optical networks use optical fiber. For this reason, we will explain how light propagates through optical fiber.

In optical communications the fiber is simply the transmission medium that transmits a light pulse generated by the optical transmitter. The presence of a pulse means the bit is worth 1, and its absence, 0.

5.1.3.1 Types of Fibers

There are three types of optical fibers depending on the diameter of the core and its refraction index. Light travels differently in each of these fiber types. The diameters of the core are measured in microns (a micron is a micrometer, which is the millionth part of a meter).

5.1.3.1.1 Step Index Multimode Fiber

In the multimode fibers, the diameter of the core varies from 50 to 100 microns. Additionally, the light rays enter through a finite number of angles and travel through a finite number of routes. The other possible routes interact with one another and are destroyed by a process of interference. They are called step index because the refraction index of the core is constant.

5.1.3.1.2 Graded Index Multimode Fiber

This fiber has the same characteristics as the step index multimode fiber with respect to the core diameter and the light rays entering from different angles. The difference between the two is that in graded index multimode fiber the core has a variable refraction index — thus the name graded index. The nearer it is to the center, the greater the refraction index of the core.

5.1.3.1.3 Step Index Monomode Fiber

In these fibers the diameter of the core goes from 7 to 9 microns, that is, it is considerably thinner than in the multimode fibers. This is possible because in these fibers the light does not bounce, but travels in a straight line, so only one light ray passes through, and there is almost no distortion.

5.2 New Optimization Functions

In the case of optical transmissions, we must again specify the meaning of each one of the variables for the unicast case as well as for the multicast one.

As explained in the case of unicast traffic for the nonoptical networks, we define the following variables:

$G = (N, E)$ is the graph that models this optical network, with N being the set of nodes and E the set of links.

F is the set of data flows that are transmitted in the network.

s and t are the source and destination flows for f flow.

$(i, j) \in E$ is the link between nodes i and j.

d_{ij} is the delay of link (i, j).

bw_f is the demand of f flow.

In the case of optical networks we must define other variables:

$\lambda \in \Lambda$ is the wavelength (λ) used, with Λ the set of values that support the network. In this case, binary variable $X_{ij}^{\lambda f}$ takes the value of 1 if the λ wavelength of link (i, j) is used to transport f flow from the s origin node up to the t destination node. If it is not used, it takes the value of 0.

P_i is the power of laser in node i.

L_{ij} is the distance of node i to node j.

A_{ij} is the attenuation factor of the fiber between nodes i and j.

bw is the bandwidth of λ.

For the multicast traffic we define the following variables:

$G = (N, E)$ is the graph that models the optical network, with N being the set of nodes and E the set of links.

F is the set of data flows that are transmitted in the network.

s is the flow source node f.

T_f is the set of nodes exiting from flow f.

$(i, j) \in E$ is the link between nodes i and j.

d_{ij} is the delay of link (i, j).

bw_f is the demand of flow f.

$\lambda \in \Lambda$ is the wavelength (λ) used, with Λ being the set of all the Λ values supported by the network. In this case, binary variable $X_{ij}^{\lambda ft}$ takes the value of 1 if the λ wavelength of link (i, j) is used by the tree to transport f flow from the s origin node up to destination node t. If it is not used, it takes the value 0, where:

P_i is the laser power in node i.

L_{ij} is the distance of node i to node j.

A_{ij} is the attenuating factor of the fiber between nodes i and j.

c_{ij} is the wavelength of λ in fiber-optical link (i, j).

In Figure 5.1 we can see that in transmissions through GMPLS the information would be transmitted through a λ, which goes through a fiber-optical link (i, j).

5.2.1 Number of λ

The first function to optimize is the quantity of λ used. This problem arises because at present, optical commutation equipment can only work with up to 128 different λ values, and this presents a restriction. Another restriction is that normally present optical commutation equipment does not carry out a change of λ. When the equipment receives a specific λ through an optical port, commutation consists in that same λ_1 being transferred through another optical port. If the optical commuter would carry out a change of λ values when this equipment receives a specific λ_1 through an optical port, it could transfer a different λ_2 through another port. Due to these characteristics, the

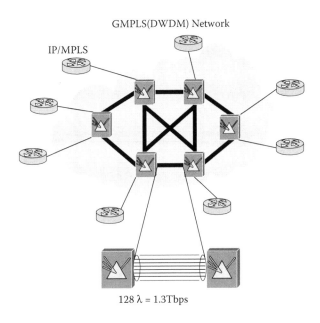

$$128 \lambda = 1.3\text{Tbps}$$

Figure 5.1 GMPLS network.

function of the number of λ used is an important factor in this type of network.

5.2.1.1 Unicast

The function may be expressed in the following manner:

$$\min \sum_{\lambda} \max(X_{ij}^{\lambda f})_{f \in F, (i,j) \in E}$$

The $\max(X_{ij}^{\lambda f})$ function allows us to determine whether a specific λ is used in a network link. The addition allows us to calculate how many λ values are used.

5.2.1.2 Multicast

For multicast transmissions, the function may be expressed in the following manner:

$$\min \sum_{\lambda} \max(X_{ij}^{\lambda ft})_{f \in F, t \in T_f, (i,j) \in E}$$

5.2.2 Optical Attenuation

Attenuation of an optical fiber link (i, j) is the loss of signal due to:

A_{ij} attenuation in the link, which is a physical characteristic of the fiber that is used

The power of transmitter P_i

The distance covered by fiber L_{ij}

The mathematical expression for the attenuation is $\left(10^{-A_{ij}*L_{ij}/10} * P_i\right)$. In this case, we want to minimize the maximum attenuation in each link of a route, from the s origin node up to the t destination node.

5.2.2.1 Unicast

The function may be expressed in the following manner:

$$\min\left\{\max\left[\left(10^{-A_{ij}*L_{ij}/10}*P_i\right)*X_{ij}^{\lambda f}\right]_{f\in F,\lambda\in\Lambda,(i,j)\in E}\right\}$$

As mentioned previously, the objective consists of minimizing the maximum attenuation of the links in a specific route. Through the max function we achieve this objective.

5.2.2.2 Multicast

For multicast transmissions, the function may be expressed in the following manner:

$$\min\left\{\max\left[\left(10^{-A_{ij}*L_{ij}/10}*P_i\right)*\max\left(X_{ij}^{\lambda ft}\right)_{t\in T_f}\right]_{f\in F,\lambda\in\Lambda,(i,j)\in E}\right\}$$

In this case, we minimize maximum attenuation for each one of the t destination nodes. As explained in Chapter 4, it is necessary to use the function $\max\left(X_{ij}^{\lambda ff}\right)$ as it is transmitting multicast flow.

5.3 Redefinition of Optical Transmission Functions

The functions that were analyzed in Chapter 4 can also be implemented in GMPLS networks. It would practically consist of adding variable λ and the corresponding additions in the functions.

5.3.1 Unicast

In the case of unicast transmissions, the functions would remain as shown in Table 5.1.

5.3.2 Multicast

In the case of multicast transmissions, the functions would remain as shown in Table 5.2.

Table 5.1 Objective Functions

	Function	Expression
f_1	Number of λ	$\min \sum_\lambda \max(X_{ij}^{\lambda f})_{f \in F, (i,j) \in E}$
f_2	Optical attenuation	$\min \left\{ \max \left[\left(10^{-A_{ij} * L_{ij}/10} * P_i \right) * X_{ij}^{\lambda f} \right]_{f \in F, \lambda \in \Lambda, (i,j) \in E} \right\}$
f_3	Hop count	$\sum_{f \in F} \sum_{\lambda \in \Lambda} \sum_{(i,j) \in E} X_{ij}^{f\lambda}$
f_4	End-to-end delay	$\sum_{f \in F} \sum_{\lambda \in \Lambda} \sum_{(i,j) \in E} d_{ij} . X_{ij}^{f\lambda}$
f_5	Cost	$\sum_{f \in F} \sum_{\lambda \in \Lambda} \sum_{(i,j) \in E} w_{ij} . X_{ij}^{f\lambda}$
f_6	Bandwidth consumption	$\sum_{f \in F} \sum_{\lambda \in \Lambda} \sum_{(i,j) \in E} bw_f . X_{ij}^{f\lambda}$
f_7	Packet loss rate	$\sum_{f \in F} \sum_{\lambda \in \Lambda} \prod_{(i,j) \in E} plr_{ij} . X_{ij}^{f\lambda}$
f_8	Blocking probability	$\max BP = \dfrac{Connection_{real}}{connection_{total}}$ or $\max Connection_{real} = \sum_{f \in F} \max \left(X_{ij}^{f} \right)_{\lambda \in \Lambda, (i,j) \in}$
f_9	Maximum link utilization	$\min \alpha$, where $\alpha = \max \left\{ \alpha_{ij} \right\}$, where $\alpha_{ij} = \dfrac{\sum_{f \in F} \sum_{\lambda \in \Lambda} bw_f . X_{ij}^{f\lambda}}{|\Lambda| . \lambda\_bw}$

Table 5.2   Objective Functions

	Function	Expression
f_1	Number of	$\min \sum_\lambda \max(X_{ij}^{\lambda ft})_{f \in F, t \in T_f, (i,j) \in E}$
f_2	Optical attenuation	$\min \left\{ \max \left[\left(10^{-A_{ij}*L_{ij}/10} * P_i \right) * \max \left(X_{ij}^{\lambda ft} \right)_{t \in T_f} \right]_{f \in F, \lambda \in \Lambda, (i,j) \in E} \right\}$
f_3	Hop count	$\min \sum_{f \in F} \sum_{t \in T_f} \sum_{\lambda \in \Lambda} \sum_{(i,j) \in E} X_{ij}^{\lambda tf}$
f_4	End-to-end delay	$\min \sum_{f \in F} \sum_{t \in T_f} \sum_{\lambda \in \Lambda} \sum_{(i,j) \in E} d_{ij}.X_{ij}^{\lambda tf}$
f_5	Cost	$\min \sum_{f \in F} \sum_{\lambda \in \Lambda} \sum_{(i,j) \in E} w_{ij}.\max \left(X_{ij}^{\lambda tf} \right)_{t \in T_f}$
f_6	Bandwidth consumption	$\min \sum_{f \in F} \sum_{\lambda \in \Lambda} \sum_{(i,j) \in E} bw_f.\max \left(X_{ij}^{\lambda tf} \right)_{t \in T_f}$
f_7	Packet loss rate	$\min \sum_{f \in F} \sum_{t \in T_f} \sum_{\lambda \in \Lambda} \prod_{(i,j) \in E} plr_{ij}.X_{ij}^{\lambda tf}$
f_8	Blocking probability	$\max BP = \dfrac{Connection_{real}}{Connection_{total}}$ or $\max Connection_{real} = \sum_{f \in F} \max \left(X_{ij}^{\lambda tf} \right)_{\lambda \in \Lambda, (i,j) \in E, t \in T_f}$
f_9	Maximum link utilization	$\min \alpha$, **where** $\alpha = \max \left\{ \alpha_{ij} \right\}$, where $\alpha_{ij} = \dfrac{\sum_{f \in F} \sum_{\lambda \in \Lambda} bw_f.\max \left(X_{ij}^{\lambda tf} \right)}{|\Lambda|.\lambda\_bw}$

5.4 Constraints

5.4.1 Unicast Transmission

In the case of transmissions with optical commutation, we must establish the following premise when a switch port receives a package through wavelength λ_1: the switch will commute and use the same value λ_1 for one of the outlet ports. This section will not analyze what happens when it is possible to make the change of λ values within the switch.

As with the transmissions without optical commutation, the first three constraints ensure that the solutions obtained are valid routes from the s origin up to the t destination.

The first constraint ensures that for each f flow there is a route from the s origin node. The equation that models this constraint is given by the following expression:

$$\sum_{\lambda \in \Lambda} \sum_{(i,j) \in E} X_{ij}^{\lambda f} = 1, f \in F, i = s$$

In Figure 5.2 we can observe that in each link (i, j) there are two wavelengths, λ_1 and λ_2. In this case, adding the outlet links of the s origin node, we realize that $X_{12}^{11} + X_{12}^{21} + X_{13}^{11} + X_{13}^{21} = 1$, and therefore, this constraint complies. It is important to remember that in this model we have added an index that identifies wavelength λ.

The second constraint ensures that for each f flow there is a route up to the t destination node. The equation that models this constraint is given by the following expression:

$$\sum_{\lambda \in \Lambda} \sum_{(j,i) \in E} X_{ji}^{\lambda f} = -1, i = t, f \in F$$

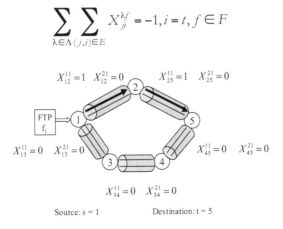

$$X_{12}^{11} = 1 \quad X_{12}^{21} = 0 \qquad X_{25}^{11} = 1 \quad X_{25}^{21} = 0$$

$$X_{13}^{11} = 0 \quad X_{13}^{21} = 0 \qquad\qquad X_{45}^{11} = 0 \quad X_{45}^{21} = 0$$

$$X_{34}^{11} = 0 \quad X_{34}^{21} = 0$$

Source: $s = 1$ Destination: $t = 5$

Figure 5.2   Unicast constraints in GMPLS networks.

In Figure 5.2, for example, adding up the arrival links for destination 5, we observe that $-X_{25}^{11} - X_{25}^{21} - X_{45}^{11} - X_{45}^{21} = -1$, and therefore, we comply with this constraint.

The third constraint ensures that for each f flow, everything that arrives to an intermediate node — in other words, that does not belong to the s origin node or to the t destination node — also exits from this intermediate node through any other link. In this case, if we add the outlets and subtract the inlets in an intermediate node, the value should be 0 for each f flow. The equation that models this constraint is

$$\sum_{\lambda \in \Lambda} \sum_{(i,j) \in E} X_{ij}^{\lambda f} - \sum_{\lambda \in \Lambda} \sum_{(j,i) \in E} X_{ji}^{\lambda f} = 0, f \in F, i \neq s, i \neq t$$

In Figure 5.2, the intermediate nodes are nodes 2, 3, and 4. For node 2, if we add its outlets and subtract its inlets, we would find that $X_{25}^{11} + X_{25}^{21}$ $- X_{12}^{11} + X_{12}^{21} = 0$. For node 3 we would find that $X_{34}^{11} + X_{34}^{21} - X_{13}^{11} + X_{13}^{21} = 0$, and, finally, for node 4 we would find that $X_{45}^{11} + X_{45}^{21} - X_{34}^{11} + X_{34}^{21} = 0$. Therefore, this constraint also complies.

On the other hand, as in the models presented, we have analyzed the traffic demand necessary for each data flow and the capacity of each λ belonging to a (i, j) fiber channel, and it is necessary to define a constraint to avoid sending a traffic demand larger than the capacity of the link. This constraint may be expressed in the following manner:

$$\sum_{f \in F} bw_f . X_{ij}^{\lambda f} \leq c_{\lambda ij}, (i, j) \in E, \lambda \in \Lambda$$

$\sum_{f \in F} bw_f . X_{ij}^{\lambda f}$ represents, in bits per second (bps), the total demand flow (for all f flows) that is transmitted by the λ wavelength in optical fiber link $(i, j) \in E$. The expression $\leq c_{\lambda ij}$ does not permit the total flow demand to be superior to the capacity of λ in link (i, j).

For optical transmissions with λ commutations, it is necessary to define a new restriction.

This new constraint limits the number of λ used in the link to the number of λ defined in network $|\Lambda|$. This constraint can be expressed in the following manner:

$$\sum_{\lambda \in \Lambda} \max\left(X_{ij}^{\lambda f}\right)_{f \in F} \leq |\Lambda|, (i,j) \in E$$

5.4.2 Multicast Transmission

In the same manner as the unicast transmissions, the first three constraints ensure that the solutions obtained are valid routes from the s origin up to each one of the $t \varepsilon T_f$ destinations.

The first constraint ensures that for the f flow there is only one route that exits from the s origin node for each $t \varepsilon T$ destination. This constraint can be expressed in the following manner:

$$\sum_{(i,j) \in E} X_{ij}^{\lambda t f} = 1, \lambda \in \Lambda, t \in T_f, f \in F, i = s$$

For example, Figure 5.3 does not include variables $X_{ij}^{\lambda t f}$ in link $(1, 3)$ with destination to node 5, because through this link it is impossible to arrive to destination node 5. In the optimization process, the value of these $X_{ij}^{\lambda t f}$ variables would be 0. This constraint complies, because adding the outlet links for origin node $s = 1$ with destination to node $t = 5$ we obtain $X_{12}^{151} + X_{12}^{251} + X_{14}^{151} + X_{14}^{251} = 1$, so this constraint of the origin node complied. In this example for destination node 6, the analysis would be similar. If we add the outlet links for origin node 1 with destination to node 6, we would see that $X_{13}^{161} + X_{13}^{261} + X_{14}^{161} + X_{14}^{261} = 1$, so this constraint of the origin node complies.

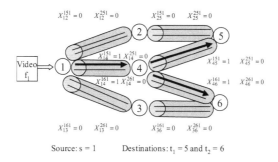

Source: s = 1   Destinations: $t_1 = 5$ and $t_2 = 6$

Figure 5.3   Multicast constraints in GMLPS networks.

The second constraint ensures that for each f flow there is only one route to each one of the t destination nodes. This constraint may be expressed in the following manner:

$$\sum_{(j,i)\in E} X_{ji}^{\lambda t f} = -1, i = t, \lambda \in \Lambda, t \in T_f, f \in F$$

In Figure 5.3, adding the arrival links for destination nodes 5 and 6 we find that $-X_{25}^{151} - X_{25}^{251} - X_{45}^{151} - X_{45}^{251} = -1$ and $-X_{36}^{161} - X_{36}^{261} - X_{46}^{161} - X_{46}^{261} = -1$, respectively. Therefore, this constraint complies.

The third constraint ensures that for each f flow everything that arrives to an intermediate node also leaves that node through any other link. This means that if we add the outlet links and subtract the inlet links in an intermediate node, the value should be 9 for each f flow. This constraint can be expressed in the following manner:

$$\sum_{(i,j)\in E} X_{ij}^{\lambda t f} - \sum_{(j,i)\in E} X_{ji}^{\lambda t f} = 0, \lambda \in \Lambda, t \in T_f, f \in F, i \neq s, i \notin T_f$$

In Figure 5.3 intermediate nodes are nodes 2, 3, and 4. For node 2, if we add the outlets and subtract the inlets, we see that $X_{25}^{151} + X_{25}^{251} - X_{12}^{151} - X_{12}^{251} = 0$. For node 3 we see that $X_{36}^{151} + X_{36}^{251} - X_{13}^{151} - X_{13}^{251} = 0$, and, finally, for node 4 we see that $X_{45}^{151} + X_{45}^{251} - X_{14}^{151} - X_{14}^{251} = 0$ and $X_{46}^{161} + X_{46}^{261} - X_{14}^{161} - X_{14}^{261} = 0$. Therefore, this constraint also complies.

In the models presented, we have analyzed the traffic demand necessary for each multicast data flow and the capacity of each λ belonging to an (i, j) optical fiber channel. However, it is necessary to define a constraint to avoid sending a larger traffic demand than the capacity of the link. This constraint can be expressed in the following manner:

$$\sum_{f\in F} bw_f . \max\left(X_{ij}^{\lambda t f}\right)_{t\in T_f} \leq c_{\lambda ij}, (i,j) \in E, \lambda \in \Lambda$$

$\sum_{f\in F} bw_f . \max\left(X_{ij}^{\lambda t f}\right)_{t\in T_f}$ represents, in bps, the total flow demand (for

all f flows) that is transmitted by the λ wavelength in the $(i, j) \in E$ optical fiber link. The expression $\leq c_{\lambda ij}$ does not allow all this flow of demand to be greater than the capacity of λ in link (i, j).

As we did in the case of unicast transmissions, we define a new constraint in the multicast transmissions. This new constraint limits the number of λ used in a link to the number of λ defined in the $|\Lambda|$ network. This constraint for the case of multicast can be expressed in the following manner:

$$\sum_{\lambda \in \Lambda} \max\left(X_{ij}^{\lambda tf}\right)_{t \in T_f, f \in F} \leq |\Lambda|, \left(i, j\right) \in E$$

5.5 Functions and Constraints

In this section we summarize the functions and constraints presented.

5.5.1 Unicast Transmissions

Table 5.3 presents a summary of the functions that were defined previously for unicast transmissions in GMPLS optical networks.

5.5.2 Multicast Transmissions

Table 5.4 presents a summary of the functions that were previously defined for the multicast transmissions in GMPLS optical networks.

5.6 Multi-objective Optimization Modeling

This section presents a real multi-objective scheme to minimize the functions number of λ, attenuation, and delay for unicast and multicast transmissions. In this case, the functions are managed as independent functions and no approximation is carried out through a single-objective method.

5.6.1 Unicast Transmissions

$$F\left(X_{ij}^{\lambda f}\right) = \left\{f_1\left(X_{ij}^{\lambda f}\right), f_2\left(X_{ij}^{\lambda f}\right), f_3\left(X_{ij}^{\lambda f}\right)\right\}$$

Table 5.3   Objective Functions

	Function	Expression
f_1	Number of λ	$\min \sum_\lambda \max(X_{ij}^{\lambda f})_{f \in F, (i,j) \in E}$
f_2	Optical attenuation	$\min \left\{ \max \left[\left(10^{-A_{ij}*L_{ij}/10} * P_i \right) * X_{ij}^{\lambda f} \right]_{f \in F, \lambda \in \Lambda, (i,j) \in E} \right\}$
f_3	Hop count	$\sum_{f \in F} \sum_{\lambda \in \Lambda} \sum_{(i,j) \in E} X_{ij}^{f\lambda}$
f_4	End-to-end delay	$\sum_{f \in F} \sum_{\lambda \in \Lambda} \sum_{(i,j) \in E} d_{ij}.X_{ij}^{f\lambda}$
f_5	Cost	$\sum_{f \in F} \sum_{\lambda \in \Lambda} \sum_{(i,j) \in E} w_{ij}.X_{ij}^{f\lambda}$
f_6	Bandwidth consumption	$\sum_{f \in F} \sum_{\lambda \in \Lambda} \sum_{(i,j) \in E} bw_f.X_{ij}^{f\lambda}$
f_7	Packet loss rate	$\sum_{f \in F} \sum_{\lambda \in \Lambda} \prod_{(i,j) \in E} plr_{ij}.X_{ij}^{f\lambda}$
f_8	Blocking probability	$\max BP = \dfrac{Connection_{real}}{Connection_{total}}$ or $\max Connection_{real} = \sum_{f \in F} \max \left(X_{ij}^{f} \right)_{\lambda \in \Lambda, (i,j) \in}$
f_9	Maximum link utilization	$\min \alpha$, where $\alpha = \max \left\{ \alpha_{ij} \right\}$, where $\alpha_{ij} = \dfrac{\sum_{f \in F} \sum_{\lambda \in \Lambda} bw_f.X_{ij}^{f\lambda}}{|\Lambda|.\lambda\_bw}$

Table 5.3   Objective Functions (continued)

Constraints			
	Name		*Expression*
c_1	Source node		$\displaystyle\sum_{\lambda\in\Lambda}\sum_{(i,j)\in E} X_{ij}^{\lambda f} = 1,\, f\in F, i=s$
c_2	Destination node		$\displaystyle\sum_{\lambda\in\Lambda}\sum_{(j,i)\in E} X_{ji}^{\lambda f} = -1,\, i=t,\, f\in F$
c_3	Intermediate node		$\displaystyle\sum_{\lambda\in\Lambda}\sum_{(i,j)\in E} X_{ij}^{\lambda f} - \sum_{\lambda\in\Lambda}\sum_{(j,i)\in E} X_{ji}^{\lambda f} = 0,\, f\in F, i\neq s, i\neq t$
c_4	Link capacity		$\displaystyle\sum_{f\in F} bw_f . X_{ij}^{\lambda f} \leq c_{\lambda ij},\, (i,j)\in E, \lambda\in\Lambda$
c_5	Number of		$\displaystyle\sum_{\lambda\in\Lambda}\max\left(X_{ij}^{\lambda f}\right)_{f\in F} \leq |\Lambda|,\, (i,j)\in E$

where

$$f_1\left(X_{ij}^{\lambda f}\right) = \sum_{\lambda}\max(X_{ij}^{\lambda f})_{f\in F,(i,j)\in E}$$

$$f_2\left(X_{ij}^{\lambda f}\right) = \left\{\max\left[\left(10^{-A_{ij}*L_{ij}/10}*P_i\right)*\max\left(X_{ij}^{\lambda f}\right)\right]_{f\in F,\lambda\in\Lambda,(i,j)\in E}\right\}$$

$$f_3\left(X_{ij}^{\lambda f}\right) = \sum_{f\in F}\sum_{\lambda\in\Lambda}\sum_{(i,j)\in E} d_{ij}.X_{ij}^{f\lambda}$$

subject to

$$\sum_{\lambda\in\Lambda}\sum_{(i,j)\in E} X_{ij}^{\lambda f} = 1,\, f\in F, i=s$$

Table 5.4   Objective Functions

	Function	Expression
f_1	Number of λ	$\min \sum_{\lambda} \max(X_{ij}^{\lambda ft})_{f \in F, t \in T_f, (i,j) \in E}$
f_2	Optical attenuation	$\min \left\{ \max \left[\left(10^{-A_{ij}*L_{ij}/10} * P_i \right) * \max \left(X_{ij}^{\lambda ft} \right)_{t \in T_f} \right]_{f \in F, \lambda \in \Lambda, (i,j) \in E} \right\}$
f_3	Hop count	$\min \sum_{f \in F} \sum_{t \in T_f} \sum_{\lambda \in \Lambda} \sum_{(i,j) \in E} X_{ij}^{\lambda tf}$
f_4	End-to-end delay	$\min \sum_{f \in F} \sum_{t \in T_f} \sum_{\lambda \in \Lambda} \sum_{(i,j) \in E} d_{ij} . X_{ij}^{\lambda tf}$
f_5	Cost	$\min \sum_{f \in F} \sum_{\lambda \in \Lambda} \sum_{(i,j) \in E} w_{ij} . \max \left(X_{ij}^{\lambda tf} \right)_{t \in T_f}$
f_6	Bandwidth consumption	$\min \sum_{f \in F} \sum_{\lambda \in \Lambda} \sum_{(i,j) \in E} bw_f . \max \left(X_{ij}^{\lambda tf} \right)_{t \in T_f}$
f_7	Packet loss rate	$\min \sum_{f \in F} \sum_{t \in T_f} \sum_{\lambda \in \Lambda} \prod_{(i,j) \in E} plr_{ij} . X_{ij}^{\lambda tf}$
f_8	Blocking probability	$\max BP = \dfrac{Connection_{real}}{Connection_{total}}$ or $\max Connection_{real} = \sum_{f \in F} \max \left(X_{ij}^{\lambda tf} \right)_{\lambda \in \Lambda, (i,j) \in E, t \in T_f}$
f_9	Maximum link utilization	$\min \alpha$, **where** $\alpha = \max \left\{ \alpha_{ij} \right\}$, where $\alpha_{ij} = \dfrac{\sum_{t \in F} \sum_{\lambda \in \Lambda} bw_f . \max \left(X_{ij}^{\lambda tf} \right)}{|\Lambda| . \lambda\_bw}$

Table 5.4   Objective Functions (continued)

Constraints		
	Name	*Expression*
c_1	Source node	$\displaystyle\sum_{(i,j)\in E} X_{ij}^{\lambda tf} = 1, \lambda \in \Lambda, t \in T_f, f \in F, i = s$
c_2	Destination node	$\displaystyle\sum_{(j,i)\in E} X_{ji}^{\lambda tf} = -1, i = t, \lambda \in \Lambda, t \in T_f, f \in F$
c_3	Intermediate node	$\displaystyle\sum_{(i,j)\in E} X_{ij}^{\lambda tf} - \sum_{(j,i)\in E} X_{ji}^{\lambda tf} = 0, \lambda \in \Lambda, t \in T_f, f \in F, i \neq s, i \notin T_f$
c_4	Link capacity	$\displaystyle\sum_{f\in F} bw_f . \max\left(X_{ij}^{\lambda tf}\right)_{t\in T_f} \leq c_{\lambda ij}, (i,j) \in E, \lambda \in \Lambda$
c_5	Number of	$\displaystyle\sum_{\lambda\in\Lambda} \max\left(X_{ij}^{\lambda tf}\right)_{t\in T_f, f\in F} \leq |\Lambda|, (i,j) \in E$

$$\sum_{\lambda\in\Lambda}\sum_{(j,i)\in E} X_{ji}^{\lambda f} = -1, i = t, f \in F$$

$$\sum_{\lambda\in\Lambda}\sum_{(i,j)\in E} X_{ij}^{\lambda f} - \sum_{\lambda\in\Lambda}\sum_{(j,i)\in E} X_{ji}^{\lambda f} = 0, f \in F, i \neq s, i \neq t$$

$$\sum_{f\in F} bw_f . X_{ij}^{\lambda f} \leq c_{\lambda ij}, (i,j) \in E, \lambda \in \Lambda$$

$$\sum_{\lambda\in\Lambda} \max\left(X_{ij}^{\lambda f}\right)_{f\in F} \leq |\Lambda|, (i,j) \in E$$

$$X_{ij}^{\lambda f} = \left\{0, 1\right\}$$

5.6.2 Multicast Transmissions

$$F\left(X_{ij}^{\lambda tf}\right) = \left\{f_1\left(X_{ij}^{\lambda tf}\right), f_2\left(X_{ij}^{\lambda tf}\right), f_3\left(X_{ij}^{\lambda tf}\right)\right\}$$

where

$$f_1\left(X_{ij}^{\lambda ft}\right) = \sum_{\lambda} \max(X_{ij}^{\lambda ft})_{f \in F, t \in T_f, (i,j) \in E}$$

$$f_2\left(X_{ij}^{\lambda ft}\right) = \left\{\max\left[\left(10^{-A_{ij}*L_{ij}/10} * P_i\right) * \max\left(X_{ij}^{\lambda ft}\right)_{t \in T_f}\right]_{f \in F, \lambda \in \Lambda, (i,j) \in E}\right\}$$

$$f_3\left(X_{ij}^{\lambda ft}\right) = \sum_{f \in F}\sum_{t \in T_f}\sum_{\lambda \in \Lambda}\sum_{(i,j) \in E} d_{ij}.X_{ij}^{\lambda tf}$$

subject to

$$\sum_{(i,j) \in E} X_{ij}^{\lambda tf} = 1, \lambda \in \Lambda, t \in T_f, f \in F, i = s$$

$$\sum_{(j,i) \in E} X_{ji}^{\lambda tf} = -1, i = t, \lambda \in \Lambda, t \in T_f, f \in F$$

$$\sum_{(i,j) \in E} X_{ij}^{\lambda tf} - \sum_{(j,i) \in E} X_{ji}^{\lambda tf} = 0, \lambda \in \Lambda, t \in T_f, f \in F, i \neq s, i \notin T_f$$

$$\sum_{f \in F} bw_f.\max\left(X_{ij}^{\lambda tf}\right)_{t \in T_f} \leq c_{\lambda ij}, (i,j) \in E, \lambda \in \Lambda$$

$$\sum_{\lambda \in \Lambda} \max\left(X_{ij}^{\lambda tf}\right)_{t \in T_f, f \in F} \leq |\Lambda|, (i,j) \in E$$

$$X_{ij}^{\lambda tf} = \left\{0,1\right\}$$

5.7 Obtaining a Solution Using Metaheuristics

In this section, by applying MOEA, we will obtain a computational solution to the problem presented previously for unicast and multicast transmissions.

For the execution of the evolutionary algorithm we must first define how the solutions will be presented. It is necessary to consider that the information flows are transmitted through the values, and these are part of the (i, j) optical fiber links.

The representation of the chromosomes has been carried out considering that we are going to minimize the functions number of , attenuation in the fiber, and end-to-end delay. To optimize other functions, it may be necessary to carry out some modifications in this representation.

5.7.1 Unicast Transmissions

Figure 5.4 shows the design of the topology to be used in the unicast case.

5.7.1.1 Codification of a Chromosome

Contrary to the chromosome represented in Chapter 4, in this case we must add the concept of the λ values in each (i, j) optical fiber link. Figure 5.5 shows a chromosome (or a solution) corresponding to a route from the s origin node up to the t origin node. This is a vector with all the nodes that comprise this route. At the end of the chromosome, we have added the λ wavelength, through which the unicast flows are transmitted. In general terms, the route represented by Figure 5.4 would be given by nodes s, n_1, n_2, ..., n_i, ..., t and the λ wavelength.

If we also optimize the bandwidth consumption function, different (λ) wavelengths and even different (i, j) optical fiber links can transport different flows. Figure 5.6 shows a chromosome in which each f flow can be transmitted through a λ or even a different link.

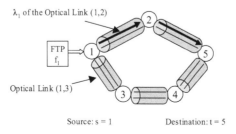

Figure 5.4   Unicast topology.

| Path | s | n₁ | n₂ | ... | nᵢ | ... | t | λ |

Represented as a chromosome with cells: Path | s | n_1 | n_2 | ... | n_i | ... | t | λ

Figure 5.5 Representation of a chromosome.

f_1	s	n_1	n_2	...	n_i	...	t	λ_1
f_2	s	n_1	n_2	...	n_i	...	t	λ_2
				...		...		
f_i	s	n_1	n_2	...	n_i	...	t	λ_i
				...		...		
$f_{|F|}$	s	n_1	n_2	...	n_i	...	t	$\lambda_{|F|}$

Figure 5.6 Representation of a chromosome.

5.7.1.2 Initial Population

To establish the initial population, we use the probabilistic breadth first search (BFS) algorithm. In this case, once we have found a route, a λ value is randomly selected for that route with $1 \leq \lambda \leq |\Lambda|$. The idea is to find several routes from the s origin node up to the t destination node to consider them as the initial population. The routes that are not found by the BFS algorithm can be found by MOEA's crossover and mutation procedures.

The probabilistic BFS algorithm will give us a population of feasible routes to resolve the problem.

5.7.1.3 Selection

This process is carried out in the same manner as the SPEA, explained previously in Chapter 4.

5.7.1.4 Crossover

This process is carried out in a manner similar to that of the SPEA, explained previously in Chapter 4. In this case, it is also necessary to determine which of the two wavelengths of the parent chromosomes is used in each of the two possible descendants. Figure 5.7 shows the function of the crossover operator.

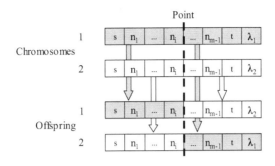

Figure 5.7   Crossover operator.

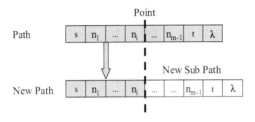

Figure 5.8   Mutation operator.

5.7.1.5 Mutation

This process is carried out in a manner similar to that of the SPEA, explained previously in Chapter 4. In this case, we maintain the same λ wavelength. Figure 5.8 shows the function of the mutation operator.

5.7.2 Multicast Transmissions

Figure 5.9 shows the design of topology used in the multicast case.

5.7.2.1 Codification of a Chromosome

In this case, we must also add the λ values through which the multicast flow is transmitted in each (i, j) fiber-optical link. The chromosome represents a tree that goes from the s origin node up to the set of $t \varepsilon T_f$ destination nodes (Figure 5.10). It is comprised of a matrix in which each file represents a route from the s origin node up to the t destination node. At the end of each file, a λ has been added, over which the flow is transmitted for that destination node. Because we have $|T_f|$ destination

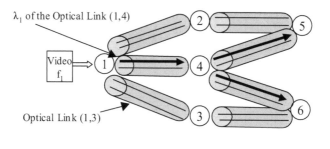

Source: s = 1   Destinations: $t_1 = 5$ and $t_2 = 6$

Figure 5.9   Multicast topology.

s	n_1	n_2	...	n_i	...	t_1	λ_1
s	n_1	n_2	...	n_i	...	t_2	λ_2
			...		...		
s	n_1	n_2	...	n_i	...	t_{Tf}	λ_{F}

Figure 5.10   Representation of chromosome 1.

nodes, the chromosome will be made up of $|T_f|$ files. In general terms, each file — in other words, each route — is represented by nodes s, n_1, n_2, ..., n_i, ..., t and the λ wavelength. This representation is valid if all the *f* flows are transmitted through the same multicast tree.

If we optimize the bandwidth consumption function, it is possible, just like we analyzed in Chapter 4, for different flows to be transmitted through different trees, through the different λ values or even through different (*i, j*) optical fibers. In this case, Figure 5.11 shows a chromosome that demonstrates that each *f* multicast flow can be transmitted through a different tree. In this case, it would be $|F|$ trees. Each T_i tree has a $|T_f|$ set of routes to carry the information to each one of the $t \epsilon T_f$ destination nodes. Finally, each P_i paid is comprised of the set of nodes from the s origin node up to one of the *t* destination nodes and the corresponding λ value for that route.

5.7.2.2 Initial Population

To establish the initial population, we use probabilistic algorithm BFS. In this case, once we have found a route, a λ value is randomly selected for that route with $1 \leq \lambda \leq |\Lambda|$. As the flow is multicast, the idea is to find

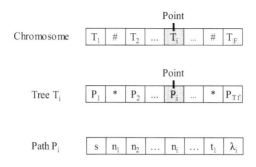

Figure 5.11   Representation of chromosome 2.

several routes from the s origin node up to each of the t destination nodes. Upon finalizing execution of this procedure, we will obtain a tree population feasible to solve the problem.

5.7.2.3 Selection

The selection process of a chromosome for the case of multicast can be carried out in the same manner as the multicast case presented in Chapter 4.

5.7.2.4 Crossover

The crossover operator depends on the representation used. If we use the chromosome represented in Figure 5.10, which comprises only one tree, the crossover functions consists of:

> Randomly selecting a crossover point
>
> Generating a first descendant, taking the first routes of the first chromosome and the last routes of the crossover point of the second chromosome
>
> Generating, if this possibility is contemplated, the second descendant, taking the second part of the first chromosome and the first part of the second

Figure 5.11 shows the proposed crossover scheme.

If we are optimizing the bandwidth consumption function and are therefore using Figure 5.12 as a representation of the chromosome, we can define the following crossover operators:

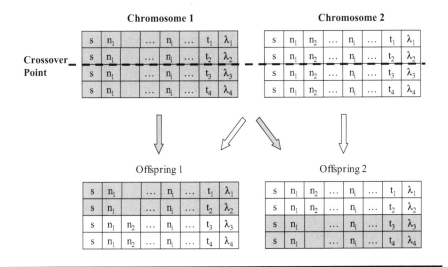

Figure 5.12   Crossover operator 1.

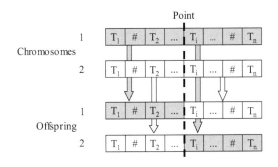

Figure 5.13   Crossover operator 2.

A first function of the crossover (Figure 5.13) consists of:
Selecting a crossover point in a random manner.

At that point taking the left side trees of the first chromosome and the right side trees of the second chromosome to generate the first offspring chromosome.

Combining for the second offspring the trees not used in the previous case.

A second crossover function for this type of chromosome consists of:

Selecting two parent chromosomes.

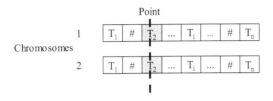

Figure 5.14 Tree selection.

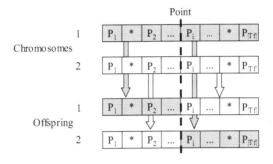

Figure 5.15 Crossover operator 3.

Randomly selecting a crossover point. In this case, we are selecting a tree for each chromosome for the same flow (Figure 5.14).

Randomly selecting a crossover point within each tree. In this case, we obtain the routes to the different flow destinations.

Combining each one of the parts of the parent chromosomes, that is, the routes of the selected tree (Figure 5.15).

Maintaining the values of λ wavelengths. Each route maintains the λ given by the parent chromosome, that is, in this function the λ values are not changed.

A third crossover function consists of:

Selecting two chromosomes.

Randomly obtaining a tree.

Randomly selecting a route in that tree (Figure 5.16).

Applying the same crossover operator as that in Section 4.7.1 (Figure 5.17). In this case, for tree T2 we have selected route P2.

Carrying out crossover between both routes.

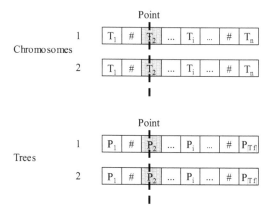

Figure 5.16   Tree and path selection.

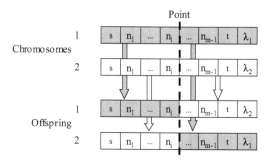

Figure 5.17   Crossover operator 4.

Assigning a λ wavelength to the offspring. It can be from the first parent, as well as from the second. In case there is a second offspring, the wavelength of the other parent must be the one assigned. For example, in Figure 5.17, to the first offspring we have assigned the λ of the second parent.

5.7.2.5 Mutation

Through the crossover operators previously defined, we have been able to carry out the necessary combinations for flows, trees, and routes. Therefore, the mutation operator can work as the operator presented in the unicast transmissions. In this case, a mutation point is selected, and

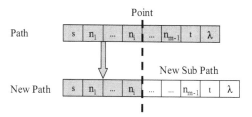

Figure 5.18   Mutation operator.

as of that mutation point, a new route is searched up to the t destination node. In this function, the λ value can be randomly changed for that route.

Figure 5.18 shows the functioning of the mutation operator.

Chapter 6

Multi-Objective Optimization in Wireless Networks

This chapter presents an extension of the optimization scheme proposed in Chapter 4 for wireless networks.

6.1 Concepts

The term *wireless* means "without wires" and includes devices capable of transmitting and receiving data without a physical connection between them, more specifically, those using radio signals. A wireless network is a system capable of connecting terminals to the data network without the need to use communication wires. Wireless networks are very useful. The most common one is the portable office, which is ideal for people with a business that requires constant traveling. Another use is that of satellite surveillance, where, for example, fleets of trucks communicate with their base station to inform about their accurate position, speed, and distance traveled during a specific period. Another important use of wireless transmissions is connectivity in metropolitan or extended networks at the microwave and satellite levels.

The availability of wireless connections and networks can broaden the users' freedom as far as the hour in which the people can solve problems related to fixed wire networks and, in some cases, even reduce networks' implementation expenses.

6.2 New Optimization Function

Wireless networks present new challenges given that data travels through the air as radio waves. Other challenges result from the unique possibilities of wireless networks, which allow freedom to move, achieved through the removal of wires. For example, the users can walk from one room to the next, from building to building, from city to city, etc., with uninterrupted wireless connection at all times.

The complexity of wireless networks increases by embedding characteristics and metrics, which are added to configuration parameters.

This chapter introduces a new parameter used for the optimization of wireless networks: free space loss.

To solve optimization problems in wireless networks, we will be using the same definition for the X_{ij}^{f} variable vector defined in the traditional unicast transmissions model of Chapter 4. The X_{ij}^{f} vector shall be defined as follows:

$$X_{ij}^{f} = \begin{cases} 1, \text{if link } (i, j) \text{ is used for flow } f \\ 0, \text{if link } (i, j) \text{ is not used for flow } f \end{cases}$$

This means that the X_{ij}^{f} variable vector is 1 if the (i, j) wireless link is used to transmit the f flow; otherwise, it is 0.

As they have been defined for unicast transmissions, the X_{ij}^{tf} variables for the case of multicast will be defined as follows:

$$X_{ij}^{tf} = \begin{cases} 1, \text{if link } (i, j) \text{ is used for flow } f \text{ with destination } t \\ 0, \text{if link } (i, j) \text{ is not used for flow } f \text{ with destination } t \end{cases}$$

This means that the X_{ij}^{tf} variable is equal to 1 if the wireless link (i, j) is used for transmitting the f multicast flow with destination node t, which belongs to the set of T_f nodes; otherwise, it is 0.

Below there is a new function related to wireless means.

6.2.1 Free Space Loss

In a communication system, free space loss is considered the attenuation suffered by the electromagnetic waves on their way from the transmitter to the receiver (Figure 6.1). This means that the transmitter sends the signal at a certain intensity and the receiver receives at a lower one.

In an ideal propagation in free space, the free space loss is the result of

Figure 6.1   Free space loss.

$$FSL = \left(\frac{4\pi d}{\lambda}\right)^2$$

where λ is the wavelength and d is the distance between the antennas.

6.2.1.1 Unicast

For unicast transmission, the function may be expressed as follows:

$$\sum_{f \in F} \max\left(FSL_{ij} * X_{ij}^f\right)_{(i,j) \in E}$$

where

$$FSL_{ij} = \left(\frac{4\pi d_{ij}}{\lambda_{ij}}\right)^2$$

In this case, the objective would consist of minimizing the maximum free space loss presented by a link along the path toward the destination node.

6.2.1.2 Multicast

For multicast transmissions, the function could be expressed as follows:

$$\sum_{f \in F} \max\left(FSL_{ij} * \max\left(X_{ij}^{tf}\right)_{t \in T_f}\right)_{(i,j) \in E}$$

where

$$FSL_{ij} = \left(\frac{4\pi d_{ij}}{\lambda_{ij}} \right)^2$$

In this case, we would be minimizing the maximum attenuation for each one of the t destination nodes.

6.3 Constraints

This section presents the basic constraints that are necessary for the solutions found to be feasible.

6.3.1 Unicast Transmission

Like the constraints presented in Chapter 4, the first three constraints ensure that the solutions obtained are feasible paths from origin s to destination t.

The first constraint makes sure that for each f flow there is only one path from origin s. The equation that models this constraint is given by the following expression:

$$\sum_{(i,j) \in E} X_{ij}^f = 1, f \in F, i = s$$

The second constraint makes sure that for each f flow there is only one path from the t destination node. The equation that models this constraint is given by

$$\sum_{(j,i) \in E} X_{ji}^f = -1, i = t, f \in F$$

The third constraint makes sure that for each f flow what arrives at an intermediate node also leaves this node from any other link. This means that if for each f flow we add the outlets and subtract the inlets in an intermediate node, the value should be 0. The equation that models this constraint is given by

$$\sum_{(i,j)\in E} X_{ij}^f - \sum_{(j,i)\in E} X_{ji}^f = 0, f \in F, i \neq s, i \neq t$$

With the above constraints we make sure that unicast transmissions in wireless networks create only one feasible path in its route from the s origin node to the t destination node.

Because in all these models we have analyzed the necessary traffic demand for each data flow and channel capacity, it is necessary to define a constraint that does not allow sending a greater traffic demand than the link is capable of receiving. The equation that models this constraint is

$$\sum_{f \in F} bw_f . X_{ij}^f \leq c_{ij}, (i,j) \in E$$

6.3.2 Multicast Transmission

As in the unicast transmissions presented in Chapter 4, the first three constraints ensure that the obtained solutions are feasible paths from the s origin to each one of the $t \varepsilon T_f$ destinations.

The first constraint makes sure that only one path comes out from the s node origin to each $t \varepsilon T_f$ destination for the f flow. The equation that models this constraint is

$$\sum_{(i,j)\in E} X_{ij}^{tf} = 1, t \in T_f, f \in F, i = s$$

The second constraint makes sure that for each f flow there is only one path going to each one of the t destinations. The equation that models this constraint is given by

$$\sum_{(j,i)\in E} X_{ji}^{tf} = -1, i = t, t \in T_f, f \in F$$

The third constraint makes sure that for each f flow, what arrives at an intermediate node also leaves this node by any other link. This means that if at an intermediate node for each f flow we add the outlets and subtract the inlets, the value should be 0. The equation that models this constraint is given by

$$\sum_{(i,j)\in E} X_{ij}^{tf} - \sum_{(j,i)\in E} X_{ji}^{tf} = 0, t \in T_f, f \in F, i \neq s, i \notin T_f$$

With the above constraints, we make sure that for multicast transmitters only one feasible path is created between the s origin node and each one of the t destination nodes.

Because in all these models we have analyzed the necessary traffic demand for each data flow and channel capacity, it is necessary to define a constraint that does not allow sending a greater traffic demand than the link is capable of receiving. The equation that models this constraint is

$$\sum_{f\in F} bw_f.\max\left(X_{ij}^{tf}\right)_{t\in T_f} \leq c_{ij}, (i,j) \in E$$

6.4 Function and Constraints

This section recapitulates only the new functions for transmissions in wireless networks and the presented constraints.

6.4.1 Unicast Transmissions

Table 6.1 is a summary of the functions defined for each unicast transmission case in wireless networks.

6.4.2 Multicast Transmissions

Table 6.2 is a summary of the functions defined for each multicast transmission case in wireless networks.

6.5 Multi-Objective Optimization Modeling

6.5.1 Unicast Transmission

If we are to optimize free space loss and delay functions, the vector for objective functions would be given by

$$F\left(X_{ij}^f\right) = \left[\sum_{f\in F}\max\left(FSL_{ij} * X_{ij}^f\right)_{(i,j)\in E}, \sum_{f\in F}\sum_{(i,j)\in E} d_{ij}.X_{ij}^f\right]$$

Table 6.1 Function Summary for Unicast Transmission

Objective Functions		
	Function	*Expression*
f_1	Free space loss	$\sum_{f\in F}\max\left(FSL_{ij}*X_{ij}^{f}\right)_{(i,j)\in E}$ where $$FSL_{ij}=\left(\frac{4\pi d_{ij}}{\lambda_{ij}}\right)^{2}$$

Constraints		
	Name	*Expression*
c_1	Source node	$\sum_{(i,j)\in E}X_{ij}^{f}=1,\,f\in F,i=s$
c_2	Destination node	$\sum_{(j,i)\in E}X_{ji}^{f}=-1,\,i=t,f\in F$
c_3	Intermediate node	$\sum_{(i,j)\in E}X_{ij}^{f}-\sum_{(j,i)\in E}X_{ji}^{f}=0\quad,f\in F,i\neq s,i\neq t$
c_4	Link capacity	$\sum_{f\in F}bw_{f}.X_{ij}^{f}\leq c_{ij}\quad,(i,j)\in E$

subject to

$$FSL_{ij}=\left(\frac{4\pi d_{ij}}{\lambda_{ij}}\right)^{2}$$

$$\sum_{(i,j)\in E}X_{ij}^{f}=1,\,f\in F,i=s$$

Table 6.2 Function Summary for Multicast Transmission

Objective Functions		
	Function	*Expression*
f_1	Free space loss	$\displaystyle\sum_{f\in F}\max\left(FSL_{ij}*\max\left(X_{ij}^{tf}\right)_{t\in T_f}\right)_{(i,j)\in E}$, where $$FSL_{ij}=\left(\frac{4\pi d_{ij}}{\lambda_{ij}}\right)^2$$

Constraints		
	Name	*Expression*
c_1	Source node	$\displaystyle\sum_{(i,j)\in E}X_{ij}^{tf}=1, t\in T_f, f\in F, i=s$
c_2	Destination node	$\displaystyle\sum_{(j,i)\in E}X_{ji}^{tf}=-1, i=t, t\in T_f, f\in F$
c_3	Intermediate node	$\displaystyle\sum_{(i,j)\in E}X_{ij}^{tf}-\sum_{(j,i)\in E}X_{ji}^{tf}=0, t\in T_f, f\in F, i\neq s, i\notin T_f$
c_4	Link capacity	$\displaystyle\sum_{f\in F}bw_f.\max\left(X_{ij}^{tf}\right)_{t\in T_f}\leq c_{ij}, (i,j)\in E$

$$\sum_{(j,i)\in E}X_{ji}^{f}=-1, i=t, f\in F$$

$$\sum_{(i,j)\in E}X_{ij}^{f}-\sum_{(j,i)\in E}X_{ji}^{f}=0, f\in F, i\neq s, i\neq t$$

$$\sum_{f \in F} bw_f . X_{ij}^f \le c_{ij} , (i,j) \in E$$

$$X_{ij}^f = \{0,1\}$$

In this case, the functions are managed as independent functions and no approach is made through a single-objective method. In other words, we are rewriting a new vector for objective functions instead of a single-objective approach. The constraints remain the same.

To conclude, we can state that the model presented in Chapter 4 can be used to carry out optimizations in wireless networks and add some additional functions that have not been anticipated in this book.

6.5.2 Multicast Transmission

If we are to optimize free space loss and delay functions in multicast transmissions, the vector of objective functions would be given by

$$\sum_{f \in F} \max \left(FSL_{ij} * \max \left(X_{ij}^{tf} \right)_{t \in T_f} \right) (i,j) \in E$$

$$F\left(X_{ij}^{tf} \right) = \left[\sum_{f \in F} \max \left(FSL_{ij} * \max \left(X_{ij}^{tf} \right)_{t \in T_f} \right)_{(i,j) \in E} , \sum_{f \in F} \sum_{t \in T_f} \sum_{(i,j) \in E} d_{ij} . X_{ij}^{tf} \right]$$

subject to

$$FSL_{ij} = \left(\frac{4\pi d_{ij}}{\lambda_{ij}} \right)^2$$

$$\sum_{(i,j) \in E} X_{ij}^{tf} = 1 , t \in T_f , f \in F , i = s$$

$$\sum_{(j,i) \in E} X_{ji}^{tf} = -1 , i = t , t \in T_f , f \in F$$

$$\sum_{(i,j)\in E} X_{ij}^{tf} - \sum_{(j,i)\in E} X_{ji}^{tf} = 0, t \in T_f, f \in F, i \neq s, i \neq t$$

$$\sum_{f\in F} bw_f \cdot \max\left(X_{ij}^{tf}\right)_{t\in T_f} \leq c_{ij}, (i,j) \in E$$

$$X_{ij}^{tf} = \{0,1\}$$

6.6 Obtaining a Solution Using Metaheuristics

The only difference between this chapter's model and the one presented in Chapter 4 is that the links (i, j) represent wireless connections. Consequently, this chapter has defined a new function related to this type of transmission means. Given that the other characteristics presented in Chapter 4 are the same as this chapter's, the designed chromosomes and operators specified in Chapter 4 work properly and can be equally used for this type of network. Therefore, if we wish to solve a problem with this new function related to wireless networks, we must refer to Chapter 4, specifically to Section 4.7, on obtaining a solution using metaheuristics.

Bibliography

[AAR89] E. Aarts. *Simulated Annealing and Boltzman Machines: A Stochastic Approach to Combinatorial Optimization and Neutral Computing*. John Wiley and Sons, New York, 1989.

[ABO98] E. Aboelela, C. Douligeris. *Fuzzy Generalized Network Approach for Solving an Optimization Model for Routing in B-ISDN*, IEEE CCECE 1998/CCGEI 1998.

[ABR02] H. Abrahamsson, B. Ahlgren, J. Alonso, A. Andersson, P. Kreuger. *A Multi Path Routing Algorithm for IP Networks Based on Flow Optimisation*, LCNS 2511. QoFIS 2002.

[AHU93] R. Ahuja, T. Magnanti, J. Orlin. *Network Flows: Theory, Algorithms and Applications*. Prentice-Hall, Englewood Cliffs, NJ, 1993.

[AHU97] R. Ahuja, M. Kodialam, A. Mishra, J. Orlin. Theory and methodology, computational investigations of maximum flow algorithms. *European Journal of Operational Research*, 97, 509–542, 1997.

[ALO02] S. Alouf, E. Altman, P. Nain. Optimal On-line Estimation of the Size of a Dynamic Multicast Group. IEEE INFOCOM 2002.

[APP03] D. Applegate, M. Thorup. Load Optimal MPLS Routing with N+M Labels. IEEE INFOCOM 2003.

[ASH02] J. Ash, M. Girish, E. Gray, B. Jamoussi, G. Wright. *Applicability Statement for CR-LDP*, RFC 3213. January 2002.

[ASH02a] J. Ash, Y. Lee, P. Ashwood-Smith, B. Jamoussi, D. Fedyk, D. Skalecki, L. Li. *LSP Modification Using CR-LDP*, RFC 3214. January 2002.

[AWD01] D. Awduche, L. Berger, D. Gan, T. Li, V. Srinivasan, G. Swallow. *RSVP-TE: Extensions to RSVP for LSP Tunnels*, RFC 3209. December 2001.

[BAC00] T. Back, D. Fogel, T. Michalewicz. *Evolutionary Computation 1*. IOP Publishing, Bristol, UK., 2000.

[BAL97] A. Ballardie. *Core Based Trees (CBT version 2) Multicast Routing*, RFC 2189. September 1997.

[BAL97a] A. Ballardie. *Core Based Trees (CBT) Multicast Routing Architecture*, RFC 2201. September 1997.

[BAN01] N. Banerjee, S. Das. Fast Determination of QoS Multicast Routes in Wireless Networks Using Genetic Algorithm. IEEE ICC 2001.

[BAR04] B. Baran, R. Fabregat, Y. Donoso, F. Solano, J.L. Marzo. Generalized Multiobjective Multitree Model, Research Report. Girona University, 2004.

[BAZ90] M. Bazaraa, J. Jarvis, H. Sherali. *Linear Programming and Network Flows*, 2nd ed. John Wiley & Sons, New York, 1990.

[BAZ93] M. Bazaraa, H. Sherali, C. Shetty. *Nonlinear Programming, Theory and Algorithms*. John Wiley & Sons, New York, 1993.

[BHA02] S. Bhatnagar, S. Ganguly, B. Nath. Label Space Reduction in Multipoint-to-Point LSPs for Traffic Engineering. ECUMN 2002.

[CAO00] Z. Cao, Z. Wang, E. Zegura. Rainbow Fair Queueing: Fair Bandwidth Sharing without Per-Flow. IEEE INFOCOM 2000.

[CAO00a] Z. Cao, Z. Wang, E. Zegura. Performance of Hashing-Based Schemes for Internet Load Balancing. IEEE INFOCOM 2000.

[CER04] S. Cerav-Erbas. Traffic Engineering in MPLS Networks with Multiple Objectives: Modeling and Optimization. Ph.D. thesis, Aachen, Germany, 2004.

[CER85] V. Cerny. Thermodynamical approach to the traveling salesman problem: an efficient simulated algorithm. *Journal of Optimization Theory and Applications*. 45, 41–45, 1985.

[CET04] C. Cetinkaya, E. Knightly. Opportunistic Traffic Scheduling over Multiple Network Paths. IEEE INFOCOM 2004.

[CHA03] D. Chakraborty, G. Chakraborty, N. Shiratori. A dynamic multicast routing satisfying multiple QoS constraints. *International Journal of Network Management*, 13:5, 321–335, 2003.

[CHE01] J. Chen, S. Chan. Multipath Routing for Video Unicast over Bandwidth-Limited Networks. IEEE GLOBECOM 2001.

[CHO03] H. Cho, J. Lee, B. Kim. Multi-path Constraint-Based Routing Algorithms for MPLS Traffic Engineering. IEEE ICC 2003.

[COE02] C. Coello, D. Van Veldhuizen, G. Lamont. *Evolutionary Algorithms for Solving Multi-objective Problems*. Kluwer Academic Publishers, New York, 2002.

[COH78] J. Cohon. *Multiobjective Programming and Planning*. Academic Press, New York, 1978.

[COL03] Y. Collette, P. Siarry. *Multiobjective Optimization, Principles and Case Studies*. Springer, Berlin, 2003.

[CRI04] J. Crichigno, B. Barán. A Multicast Routing Algorithm Using Multiobjective Optimization. IEEE ICT 2004.

[CRI04a] J. Crichigno, B. Barán. Multiobjective Multicast Routing Algorithm. IEEE ICT 2004.

[CRI04b] J. Crichigno, B. Barán. Multiobjective Multicast Routing Algorithm for Traffic Engineering. IEEE ICCCN 2004.

[CUI03] X. Cui, C. Lin, Y. Wei. A Multiobjective Model for QoS Multicast Routing Based on Genetic Algorithm. ICCNMC 2003.

[CUI03a] Y. Cui, K. Xu, J. Wu. Precomputation for Multi-constrained QoS Routing in High-Speed Networks. IEEE INFOCOM 2003.

[DEB01] K. Deb. *Multi-objective Optimization Using Evolutionary Algorithms.* John Wiley & Sons, New York, 2001.

[DEE98] S. Deering. Protocol Independent Multicast Version 2 Dense Mode. Specification, draft-ietf-pim-v2-dm-01.txt. November 1998.

[DON03] Y. Donoso, R. Fabregat, L. Fàbrega. Multi-objective Scheme over Multi-tree Routing in Multicast MPLS Networks. ACM/IFIP LANC 2003.

[DON03a] Y. Donoso, R. Fabregat. Ingeniería de Tráfico Aplicada a LSPs Punto-Multipunto en Redes MPLS. CLEI (Congreso Latinoamericano de Estudios en Informática) 2003, La Paz, Bolivia. October 2003.

[DON04] Y. Donoso, R. Fabregat, J.L. Marzo. Multi-objective Optimization Algorithm for Multicast Routing with Traffic Engineering. IEEE ICN 2004.

[DON04a] Y. Donoso, R. Fabregat, J.L. Marzo. Multi-objective Optimization Model and Heuristic Algorithm for Dynamic Multicast Routing. IEEE & VDE Networks 2004.

[DON04b] Y. Donoso, R. Fabregat, J.L. Marzo. Multi-objective Optimization Scheme for Dynamic Multicast Groups. IEEE ISCC 2004.

[DON04c] Y. Donoso, R. Fabregat, J.L. Marzo. Multi-objective optimization algorithm for multicast routing with traffic engineering. *Telecommunication Systems Journal*, 27, 229–252, 2004.

[DON04d] Y. Donoso, R. Fabregat, J.L. Marzo. Multi-objective Optimization Model and Heuristic Algorithm for Multipath Routing of Static and Dynamic Multicast Group. III Workshop on MPLS Networks, Girona, Spain. March 2004.

[DON04e] Y. Donoso, R. Fabregat, J.L. Marzo. Multicast Routing with Traffic Engineering: A Multi-objective Optimization Scheme and a Polynomial Shortest Path Tree Algorithm with Load Balancing. Proceedings of CCIO-2004, Cartagena de Indias, Colombia. March 2004.

[DON05] Y. Donoso, R. Fabregat, F. Solano, J.L. Marzo, B. Baran. Generalized Multiobjective Multitree Model for Dynamic Multicast Groups. IEEE ICC, Seoul, Korea. May 2005.

[DOR97] M. Dorigo, L. Gambardella. Ant colony system: a cooperative learning approach to the traveling salesman problem. *IEEE Transactions on Evolutionary Computation*, 1: 1, 53–66, 1997.

[DOR04] M. Dorigo, T. Sttzle. *Ant Colony Optimization.* Bradford Books, Bruxelles, Belgium, 2004.

[EST98] D. Estrin. *Protocol Independent Multicast-Sparse Mode (PIM-SM): Protocol Specification*, RFC 2362. June 1998.

[FAB04] R. Fabregat, Y. Donoso, F. Solano, J.L. Marzo. Multitree Routing for Multicast Flows: A Genetic Algorithm Approach. CCIA 2004.

[FAB04a] R. Fabregat, Y. Donoso, J.L. Marzo, A. Ariza. A Multi-objective Multipath Routing Algorithm for Multicast Flows. SPECTS 2004.

[FOR02] B. Fortz, M. Thorup. Optimizing OSPF/IS-IS weights in a changing world. *Journal on Selected Areas in Communications*, 20: 4, 756–767, 2002.

[FRA00] J. Franz, V. Jain. *Optical Communications, Components and Systems.* CRC Press, Boca Raton, FL, 2000.

[GAM04] GAMS. Solver for Large Mathematical Programming Problems, http://www.gams.com.

[GHA03] R. Ghanea-Hercock. *Applied Evolutionary Algorithms in Java.* Springer, Berlin, 2003.

[GLO97] F. Glover, M. Laguna. *Tabu Search.* Kluwer Academic Publishers, Dordrecht, The Netherlands, 1997.

[GOL89] D. Goldberg. *Genetic Algorithms in Search, Optimization and Machine Learning.* Addison-Wesley Publishing, Reading, MA, 1989.

[GUP03] A. Gupta, A. Kumar, R. Rastogi. Exploring the Trade-Off between Label Size and Stack Depth in MPLS Routing. IEEE INFOCOM 2003.

[HER03] R. Hernandez, C. Fernandez, P. Baptista. *Metodología de la Investigación,* 3rd ed. McGraw-Hill, New York, 2003.

[HOR97] J. Horn. Multicriteria decision making and evolutionary computation. In T. Baeck, D.B. Fogel, Z. Michalewicz, Eds., *Handbook of Evolutionary Computation.* Institute of Physics Publishing, Bristol, U.K., 1997.

[IM95] Y. Im, Y. Lee, S. Wi, K. Lee, Y. Choi, C. Kim. Multicast Routing Algorithms in High Speed Networks. IEEE 1995.

[INA99] J. Inagaki, M. Haseyama, H. Kitajima. A genetic algorithm for determining multiple routes and its applications. *Circuits and Systems,* 6, 137–140, 1999.

[IZM02] R. Izmailov, D. Niculescu. Flow Splitting Approach for Path Provisioning and Path Protection Problems. HPSR 2002.

[KIM02] C. Kim, Y. Choi, Y. Seok, Y. Lee. A Constrained Multipath Traffic Engineering Scheme for MPLS Networks. IEEE ICC 2002.

[KIM04] J. Kim, C. Kim, S. Seok, C. Kang. Traffic Engineering Using Adaptive Multipath-Forwarding against Dynamic Traffic in MPLS Networks. IEEE ICN 2004.

[KIR83] S. Kirkpatrick, C. Gelatt, M. Vecchi, Optimization by simulated annealing. *Science Journal,* 220, 680, 1983.

[KOY04] A. Koyama, L. Barolli, K. Matsumoto, B. Apduhan. A GA-Based Multipurpose Optimization Algorithm for QoS Routing. IEEE AINA 2004.

[LAG03] M. Laguna, R. Martí. *Scatter Search. Methodology and Implementations in C.* Kluwer Academic Publishers, Dordrecht, The Netherlands, 2003.

[LAY04] L. Layuan, L. Chunlin. QoS Multicast Routing in Networks with Uncertain Parameter. IEEE IPDPS 2004.

[LEE02] L. Lee, Y. Seok, Y. Choi, C. Kim. A Constrained Multipath Traffic Engineering Scheme for MPLS Networks. IEEE ICC 2002.

[LEU98] Y. Leung, G. Li, Z. Xu. A genetic algorithm for the multiple destination routing problems. *IEEE Transactions on Evolutionary Computation,* 2:4, 150–161, 1998.

[LI99] Y. Li, Y. Bouchebaba. *A New Genetic Algorithm for the Optimal Communication Spanning Tree Problem,* LNCS 1829. 1999.

[MED01] A. Medina, A. Lakhina, I. Mata, J. Byers. BRITE: An Approach to Universal Topology Generation. International Workshop on Modeling, Analysis and Simulation of Computer and Telecommunications Systems 2001.

[MER00] P. Merz. Memetic Algorithms for Combinatorial Optimization Problems. Ph.D. thesis, Universitat Gesamthochshule, Siegen, Germany, 2000.

[MON04] D. Montgomery. *Diseño y Análisis de Experimentos,* 2nd ed. Limusa Wiley, Mexico DF, Mexico, 2004.

[MOR80] J. Morse. Reducing the size of the nondominated set: pruning by clustering. *Computers and Operating Research*, 7(1–2), 55–66, 1980.

[MOS03] P. Moscato, C. Carlos. *Introducción a los Algortimos Meméticos*. Universidad de Málaga, Málaga, Spain, 2003.

[MOY94] J. Moy. *Multicast Extensions to OSPF*, RFC 1584. March 1994.

[OH03] Y. Oh, D. Kim, H. Yoen, M. Do, J. Lee. Scalable MPLS Multicast Using Label Aggregation in Internet Broadcasting Systems. ICT 2003.

[PAU97] A. Paulraj, C. Papadias. Space-time processing for wireless communications. *IEEE Signal Processing Magazine*, 14:b, 49–83, 1997.

[POM04] D. Pompili, L. Lopez, C. Scoglio. DIMRO: A DiffServ-Integrated Multicast Algorithm for Internet Resource Optimization in Source Specific Multicast Applications. ICC 2004.

[PUS00] T. Pusateri. Distance Vector Multicast Routing Protocol, draft-ietf-idmr-dvmrp-v3-10. August 2000.

[RAG99] S. Raghavan, G. Manimaran, C. Siva Ram Murthy. A rearrangeable algorithm for the construction of delay-constrained dynamic multicast trees. *IEEE/ACM Transactions on Networking*, 7, 514–529, 1999.

[RAO98] N. Rao, S. Batsell. QoS Routing via Multiple Paths Using Bandwidth Reservation. IEEE INFOCOM 1998.

[ROS01] E. Rosen, A. Viswanathan, R. Callon. *Multiprotocol Label Switching Architecture*, RFC 3031. 2001.

[ROY02] A. Roy, N. Banerjee, S. Das. An Efficient Multi-objective QoS-Routing Algorithm for Wireless Multicasting. IEEE INFOCOM 2002.

[ROY04] A. Roy, S. Das. QM2RP: a QoS-based mobile multicast routing protocol using multi-objective genetic algorithm. *Journal of Mobile Communication, Computation and Information, Wireless Networks*, 10:3, 271–286, 2004.

[SAI00] H. Saito, Y. Miyao, M. Yoshida. Traffic Engineering Using Multiple Multipoint-to-Point LSPs. INFOCOM 2000.

[SEO02] Y. Seok, Y. Lee, Y. Choi, C. Kim. Explicit Multicast Routing Algorithms for Constrained Traffic Engineering. IEEE ISCC 2002.

[SOL04] F. Solano, R. Fabregat, Y. Donoso. Sub-flow Assignment Model of Multicast Flows Using Multiple p2mp LSPs. 30th Conferencia Latinoamericana de Informática (CLEI2004). September 2004.

[SOL04a] F. Solano, R. Fabregat, Y. Donoso, J.L. Marzo. Mapping Subflows to P2MP LSPs. IEEE International Workshop on IP Operations and Management (IPOM 2004). Beijing, China, October 2004.

[SOL05] F. Solano, R. Fabregat, Y. Donoso, J.L. Marzo. Asymmetric Tunnels in P2MP LSPs as a Label Space Reduction Method. IEEE ICC 2005. Seoul, Korea, May 2005.

[SON03] J. Song, S. Kim, M. Lee, H. Lee, T. Suda. Adaptive Load Distribution over Multipath in MPLS Networks. IEEE ICC 2003.

[SRI03] A. Sridharan, R. Guerin, C. Diot. Achieving Near-Optimal Traffic Engineering Solutions for Current OSPF/IS-IS Networks. IEEE INFOCOM 2003.

[STR02] A. Striegel, G. Manimaran. A survey of QoS multicasting issues. *IEEE Communications Magazine*, 40:6, 82–87, 2002.

[SUN99] Q. Sun. A Genetic Algorithm for Delay-Constrained Minimum-Cost Multicasting, Technical Report, 74/75, 38106. IBR, TU Braunschweig, Butenweg, 1999.

[THA00] D. Thaler. Border Gateway Multicast Protocol (BGMP): Protocol Specification, draft-ietf-bgmp-spec-02.txt. November 2000.

[TRA03] H. Tran, R. Harris. Genetic Algorithm Approach to Rearrangement of QoS Multicast Tree. 3rd ATcrc Telecommunications and Networking Conference and Workshop. December 11–12, 2003.

[VAN89] P. Van Laarhoven, E. Aarts. *Simulated Annealing: Theory and Applications.* Kluwer Academic Publishers, Dordrecht, The Netherlands, 1989.

[VAN99] D. Van Veldhuizen. Multiobjective Evolutionary Algorithms: Classifications, Analysis and New Innovations. Ph.D. thesis, Air Force Institute of Technology, 1999.

[VIL99] C. Villamizar. MPLS Optimized Multipath (MPLS-OMP). Internet Draft. 1999.

[VUT00] S. Vutukury, J. Garcia. A Traffic Engineering Approach Based on Minimum-Delay Routing. IEEE ICCN 2000.

[XIA99] F. Xiang, L. Junzhou, W. Jieyi, G. Guanqun. QoS routing based on genetic algorithm. *Computer Communications*, 22, 1293–1399, 1999.

[WAI88] D. Waitzman. *Distance Vector Multicast Routing Protocol*, RFC 1075. 1988.

[WAN01] Z. Wang. *Internet QoS. Arquitecture and Mechanism for Quality of Service.* Morgan Kaufmann Publishers, San Francisco, 2001.

[WAN01a] Y. Wang, Z. Wang, L. Zhang. Internet Traffic Engineering without Full Mesh Overlaying. INFOCOM 2001.

[ZIT99] E. Zitzler, L. Thiele. Multiobjective evolutionary algorithm: a comparative case study and the strength Pareto approach. *IEEE Transactions on Evolutionary Computation*, 3:4, 257–271, 1999.

Annex A[1]

File in GAMS of Section 2.2.1

```
**************************************************
***        Multi-Objective Optimization        ***
***                                            ***
***      Yezid E. Donoso Meisel                ***
**************************************************
Variables
    x        Value of x
    f        minimization
    f1       function 1
    f2       function 2
    w1       Multiplication factor f1
    w2       Multiplication factor f2
;
Positive Variable w1;
Positive Variable w2;
Equations
optimal        Objective function
constraint1    constraint
constraint2    constraint
weight         Weight constraint
value_w1       value w1
value_w2       value w2
```

---
[1] These source files are available online at http://www.crcpress.com/e_products/ downloads/default.asp

```
value_f1        value f1
value_f2        value f2
;

optimal  ..   f =e= (w1*power(x,4)) + (w2*power(x-2,4));

constraint1 ..   x =l= 4;
constraint2 ..   x =g= -4;

weight   ..   w1 + w2 =e= 1;

value_w1  ..   w1 =e= 1.0;
value_w2  ..   w2 =e= 0.0;

value_f1  ..   f1 =e= power(x,4);
value_f2  ..   f2 =e= power(x-2,4);

Model Transport /all/ ;

*option dnlp=CONOPT
*option dnlp=MINOS
option nlp=SNOPT
*option lp=CPLEX
*option iterlim = 50000
*option profiletol = 1.0E-2
Solve transport using nlp minimizing f ;

Display x.l, x.m
Display f.l, f.m
Display w1.l, w1.m
Display w2.l, w2.m
Display f1.l, f1.m
Display f2.l, f2.m
```

File in GAMS of Section 2.2.2

```
****************************************************
***         Multi-Objective Optimization        ***
***                                              ***
***      Yezid E. Donoso Meisel                  ***
****************************************************
Variables
```

```
   x               Value of x
   f               minimization
   f1              function 1
   f2              function 2
   z1              feasible value f1
   z2              feasible value f2
;
Equations
optimal            Objective function

restriction_f1  restriction function 1
restriction_f2  restriction function 2
value_z1           value z1
value_z2           value z2

value_f1           value f1
value_f2           vallor f2


;

optimal   ..   f =e=  (max(0,(z1-power(x,4)))+max(0,(z2-
power(x-2,4)))));

restriccion_f1          ..   power(x,4) =l= z1;
restriccion_f2          ..   power(x-2,4) =l= z2;

value_z1                ..   z1 =e= 0.4;
value_z2                ..   z2 =e= 3;


value_f1                ..   f1 =e= power(x,4);
value_f2                ..   f2 =e= power(x-2,4);


Model Transport /all/ ;


*option  dnlp=CONOPT
*option  dnlp=MINOS
option nlp=SNOPT
```

```
*option lp=CPLEX
*option iterlim = 50000
*option profiletol = 1.0E-2
Solve transport using nlp maximizing f ;


Display x.l, x.m
Display f.l, f.m
Display f1.l, f1.m
Display f2.l, f2.m
Display z1.l, z1.m
Display z2.l, z2.m
```

File in GAMS of Section 2.2.4

```
*****************************************************
***        Multi-Objective Optimization       ***
**                                            ***
***      Yezid E. Donoso Meisel               ***
*****************************************************
Variables
   x              Value of x
   f              minimization
   f1             function 1
   f2             function 2
   w1             Multiplication factor f1
   w2             Multiplication factor f2
   z1             Reference Point f1
   z2             Reference Point f2
;
Positive Variable w1;
Positive Variable w2;
Equations
optimal           Objective function

weight            Weight constraint
value_w1          value w1
value_w2          value w2
value_z1          value z1
```

```
value_z2            value z2
value_f1            value f1
value_f2            value f2


;

optimal             ..  f =e= max((w1*abs(z1-power(x,4))),
(w2*abs(z2-power(x-2,4))));

weight              ..  w1 + w2 =e= 1;

value_w1            ..  w1 =e= 1.0;
value_w2            ..  w2 =e= 0.0;

value_z1            ..  z1 =e= 0.0;
value_z2            ..  z2 =e= 0.0;



value_f1            ..  f1 =e= power(x,4);
value_f2            ..  f2 =e= power(x-2,4);

Model Transport /all/ ;



*option dnlp=CONOPT
*option dnlp=MINOS
option nlp=SNOPT
*option lp=CPLEX
*option iterlim = 50000
*option profiletol = 1.0E-2
Solve transport using nlp minimizing f ;


Display x.l, x.m
Display f.l, f.m
Display w1.l, w1.m
Display w2.l, w2.m
Display f1.l, f1.m
Display f2.l, f2.m
```

File in GAMS of Section 2.2.5

```
************************************************
***          Multi-Objective Optimization        ***
***                                              ***
***       Yezid E. Donoso Meisel                 ***
************************************************
Variables
   x                Value of x
   f                minimization
   f1               function 1
   f2               function 2
   z1               feasible value f1
   z2               feasible value f2
;
Equations
optimal              Objective function

restriction_f1   restriction function 1
restriction_f2   restriction function 2
value_z1             value z1
value_z2             value z2

value_f1             value f1
value_f2             value f2


;

optimal                  ..   f =e= (max(0,(z1-
power(x,4)))+max(0,(z2-power(x-2,4)))));

restriction_f1     ..   power(x,4)  =l= z1;
restriction_f2     ..   power(x-2,4) =l= z2;

value_z1               ..   z1 =e= 0.4;
value_z2               ..   z2 =e= 3;

value_f1               ..   f1 =e= power(x,4);
value_f2               ..   f2 =e= power(x-2,4);

Model Transport /all/ ;
```

```
*option dnlp=CONOPT
*option dnlp=MINOS
option nlp=SNOPT
*option lp=CPLEX
*option iterlim = 50000
*option profiletol = 1.0E-2
Solve transport using nlp maximizing f ;



Display x.l, x.m
Display f.l, f.m
Display f1.l, f1.m
Display f2.l, f2.m
Display z1.l, z1.m
Display z2.l, z2.m
```

File in GAMS of Section 4.5.1.1

```
****************************************************
***                                            ***
***          Multi-Objective Ej01              ***
***                                            ***
***     Objective Functions:                   ***
***         Hop Count                          ***
***         Delay                              ***
***                                            ***
***      Eng. Yezid E. Donoso Meisel, Ph.D.    ***
****************************************************
Sets
   i    ingress node / n1, n2, n3, n4, n5 /
   j    egress node  / n1, n2, n3, n4, n5 /
   f    Flows / f1 / ;
Parameter
   bw(f) Transmission rate (Kbps)
         / f1 32 / ;
Table delay(i,j) Link Delay
                 n1      n2      n3      n4      n5
n1               999     5       3       999     20
n2               999     999     999     999     6
```

```
n3                  999     999     999     2       999
n4                  999     999     999     999     2
n5                  999     999     999     999     999 ;
Parameter d(i,j) delay ;
  d(i,j) = delay(i,j) ;
Table cap(i,j) Link Capacity (Kbps)
                    n1      n2      n3      n4      n5
n1                  0       64      64      0       16
n2                  0       0       0       0       64
n3                  0       0       0       64      0
n4                  0       0       0       0       64
n5                  0       0       0       0       0   ;
Parameter c(i,j) Link Capacity ;
  c(i,j) = cap(i,j) ;
Table cost(i,j) Link Cost
                    n1      n2      n3      n4      n5
n1                  999     5       3       999     20
n2                  999     999     999     999     6
n3                  999     999     999     2       999
n4                  999     999     999     999     2
n5                  999     999     999     999     999 ;
Parameter w(i,j) delay ;
  w(i,j) = cost(i,j) ;
Variables
  x(i,j,f) if the link i j is usd (0) to the flow f
  HC              Hop Count (Hops)
  DL              Delay (ms)
  CT              Cost ($)
  BC              Bandwidth Consumption (Kbps)
  z               minimization ;
Binary Variable x;
Equations
Shortest_Path        Objective function
source_node(f)       source node
destination_node(f)  destination node
intermediate1_node(f) intermediate node
intermediate2_node(f) intermediate node
intermediate3_node(f) intermediate node
Link_capacity(i,j)   Link Capacity ;
```

```
Shortest_Path           .. z =e= (0*(sum((f),sum((i,j),
x(i,j,f)))))
                                      +
                                  (1*(sum((f),sum((i,j),
d(i,j)*x(i,j,f)))))  ;


source_node(f)          .. sum((j), x('n1',j,f)) =e= 1;

destination_node(f)     .. sum((i),x(i,'n5',f)) =e= 1;


intermediate1_node(f) .. sum((i,j), x(i,'n2',f)) -
sum((i,j), x('n2',j,f)) =e= 0;
intermediate2_node(f) .. sum((i,j), x(i,'n3',f)) -
sum((i,j), x('n3',j,f)) =e= 0;
intermediate3_node(f) .. sum((i,j), x(i,'n4',f)) -
sum((i,j), x('n4',j,f)) =e= 0;


Link_Capacity(i,j) .. sum((f),bw(f)*x(i,j,f)) =l= c(i,j);


Model Transport /all/ ;


option mip=CPLEX
*option iterlim = 50000
*option profiletol = 1.0E-2
Solve transport using mip minimizing z ;


DL.l = sum((f),sum((i,j), d(i,j)*x.l(i,j,f)));
HC.l = sum((f),sum((i,j), x.l(i,j,f)));
CT.l = sum((f),sum((i,j), w(i,j)*x.l(i,j,f)));
BC.l = sum((f),sum((i,j), bw(f)*x.l(i,j,f)));


Display x.l
Display DL.l
Display HC.l
Display CT.l
Display BC.l
Display z.l
```

File in GAMS of Section 4.5.1.2 (f_2)

```
*****************************************************
***                                               ***
***         Multi-Objective Ej01                  ***
***                                               ***
***         E-Constraint                          ***
***                                               ***
***         Objective Functions:                  ***
***         Hop Count                             ***
***         Delay                                 ***
***                                               ***
***     Eng. Yezid E. Donoso Meisel, Ph.D.        ***
*****************************************************
Sets
   i    ingress node / n1, n2, n3, n4, n5 /
   j    egress node  / n1, n2, n3, n4, n5 /
   f    Flows / f1 / ;
Parameter
   bw(f) Transmission rate (Kbps)
         / f1 32 / ;
Table delay(i,j) Link Delay
               n1        n2        n3        n4        n5
n1             999       5         3         999       20
n2             999       999       999       999       6
n3             999       999       999       2         999
n4             999       999       999       999       2
n5             999       999       999       999       999 ;
Parameter d(i,j) delay ;
   d(i,j) = delay(i,j) ;
Table cap(i,j) Link Capacity (Kbps)
               n1        n2        n3        n4        n5
n1             0         64        64        0         16
n2             0         0         0         0         64
n3             0         0         0         64        0
n4             0         0         0         0         64
n5             0         0         0         0         0   ;
Parameter c(i,j) Link Capacity ;
   c(i,j) = cap(i,j) ;
```

```
Table cost(i,j) Link Cost
                n1        n2        n3        n4        n5
n1              999       5         3         999       20
n2              999       999       999       999       6
n3              999       999       999       2         999
n4              999       999       999       999       2
n5              999       999       999       999       999 ;
Parameter w(i,j) delay ;
  w(i,j) = cost(i,j) ;
Variables
  x(i,j,f)      if the link i j is used (0) to the flow f
  HC            Hop Count (Hops)
  DL            Delay (ms)
  CT            Cost ($)
  BC            Bandwidth Consumption (Kbps)
  z             minimization ;
Binary Variable x;
Equations
Shortest_Path         Objective function
e_constraint_f2       constraint function f2
source_node(f)        source node
destination_node(f)   destination node
intermediate1_node(f) intermediate node
intermediate2_node(f) intermediate node
intermediate3_node(f) intermediate node
Link_capacity(i,j)    Link Capacity ;

Shortest_Path .. z =e= sum((f),sum((i,j), x(i,j,f)));

e_constraint_f2 .. sum((f),sum((i,j), d(i,j)*x(i,j,f)))
=l= 15;

source_node(f) .. sum((j), x('n1',j,f)) =e= 1;

destination_node(f) .. sum((i),x(i,'n5',f)) =e= 1;

intermediate1_node(f) .. sum((i,j), x(i,'n2',f)) -
sum((i,j), x('n2',j,f)) =e= 0;
intermediate2_node(f) .. sum((i,j), x(i,'n3',f)) -
sum((i,j), x('n3',j,f)) =e= 0;
```

```
intermediate3_node(f) .. sum((i,j), x(i,'n4',f)) -
sum((i,j), x('n4',j,f)) =e= 0;


Link_Capacity(i,j) .. sum((f),bw(f)*x(i,j,f)) =l= c(i,j);



Model Transport /all/ ;

option mip=CPLEX
*option iterlim = 50000
*option profiletol = 1.0E-2
Solve transport using mip minimizing z ;

DL.l = sum((f),sum((i,j), d(i,j)*x.l(i,j,f)));
HC.l = sum((f),sum((i,j), x.l(i,j,f)));
CT.l = sum((f),sum((i,j), w(i,j)*x.l(i,j,f)));
BC.l = sum((f),sum((i,j), bw(f)*x.l(i,j,f)));

Display x.l
Display DL.l
Display HC.l
Display CT.l
Display BC.l
Display z.l
```

File in GAMS of Section 4.5.1.2 (f_1)

```
*********************************************************
***                                                 ***
***        Multi-Objective Ej01                     ***
***                                                 ***
***        E-Constraint Function f2                 ***
***                                                 ***
***        Objective Functions:                     ***
***        Hop Count                                ***
***        Delay                                    ***
***                                                 ***
***     Eng. Yezid E. Donoso Meisel, Ph.D.          ***
*********************************************************
```

```
Sets
   i     ingress node / n1, n2, n3, n4, n5 /
   j     egress node  / n1, n2, n3, n4, n5 /
   f     Flows / f1 / ;
Parameter
   bw(f) Transmission rate (Kbps)
         / f1 32 / ;
Table delay(i,j) Link Delay
                  n1      n2      n3      n4      n5
n1                999     5       3       999     20
n2                999     999     999     999     6
n3                999     999     999     2       999
n4                999     999     999     999     2
n5                999     999     999     999     999 ;
Parameter d(i,j) delay ;
   d(i,j) = delay(i,j) ;
Table cap(i,j) Link Capacity (Kbps)
                  n1      n2      n3      n4      n5
n1                0       64      64      0       16
n2                0       0       0       0       64
n3                0       0       0       64      0
n4                0       0       0       0       64
n5                0       0       0       0       0   ;
Parameter c(i,j) Link Capacity ;
   c(i,j) = cap(i,j) ;
Table cost(i,j) Link Cost
                  n1      n2      n3      n4      n5
n1                999     5       3       999     20
n2                999     999     999     999     6
n3                999     999     999     2       999
n4                999     999     999     999     2
n5                999     999     999     999     999 ;
Parameter w(i,j) delay ;
   w(i,j) = cost(i,j) ;
Variables
   x(i,j,f) if the link i j is used (0) to the flow f
   HC          Hop Count (Hops)
   DL          Delay (ms)
   CT          Cost ($)
   BC          Bandwidth Consumption (Kbps)
```

```
      z               minimizacion ;
Binary Variable x;
Equations
Shortest_Path         Objective function
e_constraint_f1       constraint function f2
source_node(f)        source node
destination_node(f)   destination node
intermediate1_node(f) intermediate node
intermediate2_node(f) intermediate node
intermediate3_node(f) intermediate node
Link_capacity(i,j)    Link Capacity ;


Shortest_Path            ..     z =e=
sum((f),sum((i,j),d(i,j)*x(i,j,f)));


e_constraint_f1          ..  sum((f),sum((i,j), x(i,j,f)))
=l= 2;


source_node(f)                 ..     sum((j), x('n1',j,f))
=e= 1;


destination_node(f)            ..     sum((i),x(i,'n5',f))
=e= 1;


intermediate1_node(f)      ..    sum((i,j), x(i,'n2',f))
- sum((i,j), x('n2',j,f)) =e= 0;
intermediate2_node(f)      ..    sum((i,j), x(i,'n3',f))
- sum((i,j), x('n3',j,f)) =e= 0;
intermediate3_node(f)      ..    sum((i,j), x(i,'n4',f))
- sum((i,j), x('n4',j,f)) =e= 0;


Link_Capacity(i,j) .. sum((f),bw(f)*x(i,j,f)) =l= c(i,j);



Model Transport /all/ ;


option mip=CPLEX
*option iterlim = 50000
*option profiletol = 1.0E-2
Solve transport using mip minimizing z ;
```

```
DL.l = sum((f),sum((i,j), d(i,j)*x.l(i,j,f)));
HC.l = sum((f),sum((i,j), x.l(i,j,f)));
CT.l = sum((f),sum((i,j), w(i,j)*x.l(i,j,f)));
BC.l = sum((f),sum((i,j), bw(f)*x.l(i,j,f)));

Display x.l
Display DL.l
Display HC.l
Display CT.l
Display BC.l
Display z.l
```

File in GAMS of Section 4.5.1.3 (*r* = 2)

```
******************************************************
***                                              ***
***         Multi-Objective Ej01                  ***
***                                              ***
***         Metric Weight  r=2                    ***
***                                              ***
***      Objective  Functions:                    ***
***      Hop  Count                               ***
***      Delay                                    ***
***                                              ***
***      Eng.  Yezid  E.  Donoso  Meisel,  Ph.D.   ***
******************************************************
Sets
  i   ingress node / n1, n2, n3, n4, n5 /
  j   egress node  / n1, n2, n3, n4, n5 /
  f   Flows / f1 / ;
Parameter
  bw(f) Transmission rate (Kbps)
        / f1 32 / ;
Table delay(i,j) Link Delay
              n1       n2       n3       n4       n5
n1            999      5        3        999      20
n2            999      999      999      999      6
```

```
n3                   999     999     999     2       999
n4                   999     999     999     999     2
n5                   999     999     999     999     999 ;
Parameter d(i,j) delay ;
   d(i,j) = delay(i,j) ;
Table cap(i,j) Link Capacity (Kbps)
                     n1      n2      n3      n4      n5
n1                   0       64      64      0       16
n2                   0       0       0       0       64
n3                   0       0       0       64      0
n4                   0       0       0       0       64
n5                   0       0       0       0       0   ;
Parameter c(i,j) Link Capacity ;
   c(i,j) = cap(i,j) ;
Table cost(i,j) Link Cost
                     n1      n2      n3      n4      n5
n1                   999     5       3       999     20
n2                   999     999     999     999     6
n3                   999     999     999     2       999
n4                   999     999     999     999     2
n5                   999     999     999     999     999 ;
Parameter w(i,j) delay ;
   w(i,j) = cost(i,j) ;
Variables
   x(i,j,f)     If the link i j is used (0) to the flow f
   HC           Hop Count (Hops)
   DL           Delay (ms)
   CT           Cost ($)
   BC           Bandwidth Consumption (Kbps)
   z            minimization ;
Binary Variable x;
Equations
Shortest_Path        Objective function
source_node(f)           source node
destination_node(f)          destination node
intermediate1_node(f)    intermediate node
intermediate2_node(f)    intermediate node
intermediate3_node(f)    intermediate node
Link_capacity(i,j)   Link Capacity ;
```

```
Shortest_Path            ..         z =e= sqrt((0.5*sqr(abs(3-
(sum((f),sum((i,j),  x(i,j,f)))))))
                                                        +
                                         (0.5*sqr(abs(3-
(sum((f),sum((i,j),  d(i,j)*x(i,j,f))))))))   ;


source_node(f)  .. sum((j),  x('n1',j,f)) =e= 1;

destination_node(f)  .. sum((i),x(i,'n5',f)) =e= 1;

intermediate1_node(f)  .. sum((i,j),  x(i,'n2',f)) -
sum((i,j),  x('n2',j,f)) =e= 0;
intermediate2_node(f)  .. sum((i,j),  x(i,'n3',f)) -
sum((i,j),  x('n3',j,f)) =e= 0;
intermediate3_node(f)  .. sum((i,j),  x(i,'n4',f)) -
sum((i,j),  x('n4',j,f)) =e= 0;

Link_Capacity(i,j)  .. sum((f),bw(f)*x(i,j,f)) =l= c(i,j);


Model Transport /all/ ;

option minlp=SBB
*option iterlim = 50000
*option profiletol = 1.0E-2
Solve transport using minlp minimizing z ;

DL.l = sum((f),sum((i,j),  d(i,j)*x.l(i,j,f)));
HC.l = sum((f),sum((i,j),  x.l(i,j,f)));
CT.l = sum((f),sum((i,j),  w(i,j)*x.l(i,j,f)));
BC.l = sum((f),sum((i,j),  bw(f)*x.l(i,j,f)));

Display x.l
Display DL.l
Display HC.l
Display CT.l
Display BC.l
Display z.l
```

File in GAMS of Section 4.5.1.3 ($r = \infty$)

```
****************************************************
***                                              ***
***          Multi-Objective Ej01                ***
***                                              ***
***          Metric Weight  r=INF                ***
***                                              ***
***          Objective  Functions:               ***
***          Hop Count                           ***
***          Delay                               ***
***                                              ***
***      Eng.  Yezid E.  Donoso Meisel,  Ph.D.   ***
****************************************************
Sets
   i    ingress node / n1, n2, n3, n4, n5 /
   j    egress node  / n1, n2, n3, n4, n5 /
   f    Flows / f1 / ;
Parameter
   bw(f) Transmission rate (Kbps)
         / f1 32 / ;
Table delay(i,j)  Link Delay
                 n1       n2       n3       n4       n5
   n1            999      5        3        999      20
   n2            999      999      999      999      6
   n3            999      999      999      2        999
   n4            999      999      999      999      2
   n5            999      999      999      999      999  ;
Parameter d(i,j)  delay ;
   d(i,j) = delay(i,j) ;
Table cap(i,j)  Link Capacity (Kbps)
                 n1       n2       n3       n4       n5
   n1            0        64       64       0        16
   n2            0        0        0        0        64
   n3            0        0        0        64       0
   n4            0        0        0        0        64
   n5            0        0        0        0        0    ;
Parameter c(i,j)  Link Capacity ;
   c(i,j) = cap(i,j) ;
```

```
Table cost(i,j) Link Cost
                n1       n2       n3       n4       n5
n1              999      5        3        999      20
n2              999      999      999      999      6
n3              999      999      999      2        999
n4              999      999      999      999      2
n5              999      999      999      999      999 ;
Parameter w(i,j) delay ;
   w(i,j) = cost(i,j) ;
Variables
   x(i,j,f)    If the link i j is used (0) to the flow f
   HC          Hop Count (Hops)
   DL          Delay (ms)
   CT          Cost ($)
   BC          Bandwidth Consumption (Kbps)
   z           minimization ;
Binary Variable x;
Equations
Shortest_Path            Objective function
source_node(f)           source node
destination_node(f)      destination node
intermediate1_node(f)    intermediate node
intermediate2_node(f)    intermediate node
intermediate3_node(f)    intermediate node
Link_capacity(i,j)       Link Capacity ;

Shortest_Path            ..           z =e= max((0.9*abs(1-
(sum((f),sum((i,j),  x(i,j,f))))))),
                                (0.1*abs(20-(sum((f),sum((i,j),
d(i,j)*x(i,j,f))))))))   ;


source_node(f)  .. sum((j),  x('n1',j,f))  =e=  1;

destination_node(f)  .. sum((i),x(i,'n5',f))  =e=  1;

intermediate1_node(f)  .. sum((i,j),  x(i,'n2',f)) -
sum((i,j),  x('n2',j,f)) =e=  0;
intermediate2_node(f)  .. sum((i,j),  x(i,'n3',f)) -
sum((i,j),  x('n3',j,f)) =e=  0;
```

```
intermediate3_node(f)  .. sum((i,j),  x(i,'n4',f)) -
sum((i,j),  x('n4',j,f)) =e=  0;


Link_Capacity(i,j)  .. sum((f),bw(f)*x(i,j,f)) =l= c(i,j);


Model Transport /all/ ;

option minlp=SBB
*option iterlim = 50000
*option profiletol = 1.0E-2
Solve transport using minlp minimizing z ;

DL.l = sum((f),sum((i,j),  d(i,j)*x.l(i,j,f)));
HC.l = sum((f),sum((i,j),  x.l(i,j,f)));
CT.l = sum((f),sum((i,j),  w(i,j)*x.l(i,j,f)));
BC.l = sum((f),sum((i,j),  bw(f)*x.l(i,j,f)));

Display x.l
Display DL.l
Display HC.l
Display CT.l
Display BC.l
Display z.l
```

File in GAMS of Section 4.5.1.4

```
********************************************************
***                                                  ***
***        Multi-Objective Ej01                      ***
***                                                  ***
***          Benson                                  ***
***                                                  ***
***        Objective  Functions:                     ***
***        Hop Count                                 ***
***        Delay                                     ***
***                                                  ***
***     Eng. Yezid E. Donoso Meisel,  Ph.D.          ***
********************************************************
```

```
Sets
   i     ingress node / n1, n2, n3, n4, n5 /
   j     egress node  / n1, n2, n3, n4, n5 /
   f     Flows / f1 / ;
Parameter
   bw(f) Transmission rate (Kbps)
         / f1 32 / ;
Table delay(i,j)  Link Delay
                n1        n2        n3        n4        n5
n1              999       5         3         999       20
n2              999       999       999       999       6
n3              999       999       999       2         999
n4              999       999       999       999       2
n5              999       999       999       999       999 ;
Parameter  d(i,j) delay ;
   d(i,j) = delay(i,j) ;
Table cap(i,j)  Link Capacity  (Kbps)
                n1        n2        n3        n4        n5
n1              0         64        64        0         16
n2              0         0         0         0         64
n3              0         0         0         64        0
n4              0         0         0         0         64
n5              0         0         0         0         0   ;
Parameter  c(i,j) Link Capacity ;
   c(i,j) = cap(i,j) ;
Table cost(i,j)  Link Cost
                n1        n2        n3        n4        n5
n1              999       5         3         999       20
n2              999       999       999       999       6
n3              999       999       999       2         999
n4              999       999       999       999       2
n5              999       999       999       999       999 ;
Parameter  w(i,j) delay ;
   w(i,j) = cost(i,j) ;
Variables
   x(i,j,f)     If the link i j is used (0) to the flow f
   HC           Hop Count (Hops)
   DL           Delay (ms)
   CT           Cost ($)
   BC           Bandwidth Consumption (Kbps)
```

```
      z              minimizacion  ;
Binary Variable x;
Equations
Shortest_Path            Objective function
e_constraint_f1          constraint function f1
e_constraint_f2          constraint function f2
source_node(f)           source node
destination_node(f)      destination node
intermediate1_node(f)    intermediate node
intermediate2_node(f)    intermediate node
intermediate3_node(f)    intermediate node
Link_capacity(i,j)       Link Capacity ;

Shortest_Path       ..  z =e= max(0,(3-(sum((f),sum((i,j),
x(i,j,f))))))

                    +
                            max(0,(7-(sum((f),sum((i,j),
d(i,j)*x(i,j,f))))))  ;

e_constraint_f1 .. sum((f),sum((i,j), x(i,j,f))) =l= 4;
e_constraint_f2 .. sum((f),sum((i,j), d(i,j)*x(i,j,f)))
=l= 10;

source_node(f)      .. sum((j), x('n1',j,f)) =e= 1;

destination_node(f) .. sum((i),x(i,'n5',f)) =e= 1;

intermediate1_node(f).. sum((i,j), x(i,'n2',f)) -
sum((i,j), x('n2',j,f)) =e= 0;
intermediate2_node(f) .. sum((i,j), x(i,'n3',f)) -
sum((i,j), x('n3',j,f)) =e= 0;
intermediate3_node(f) .. sum((i,j), x(i,'n4',f)) -
sum((i,j), x('n4',j,f)) =e= 0;

Link_Capacity(i,j) .. sum((f),bw(f)*x(i,j,f)) =l= c(i,j);


Model Transport /all/ ;

option minlp=SBB
*option iterlim = 50000
```

```
*option profiletol = 1.0E-2
Solve transport using minlp maximizing z ;

DL.l = sum((f),sum((i,j), d(i,j)*x.l(i,j,f)));
HC.l = sum((f),sum((i,j), x.l(i,j,f)));
CT.l = sum((f),sum((i,j), w(i,j)*x.l(i,j,f)));
BC.l = sum((f),sum((i,j), bw(f)*x.l(i,j,f)));

Display x.l
Display DL.l
Display HC.l
Display CT.l
Display BC.l
Display z.l
```

File in GAMS of Section 4.5.2.1

```
*****************************************************
***                                             ***
***         Multi-Objective Ej01                ***
***                                             ***
***       Objective  Functions:                 ***
***       Hop Count                             ***
***       Delay                                 ***
***                                             ***
***     Eng. Yezid E. Donoso Meisel, Ph.D.      ***
*****************************************************
Sets
   i   ingress node / n1, n2, n3, n4, n5, n6 /
   j   egress node  / n1, n2, n3, n4, n5, n6 /
   f   Flows / f1 /
   t   destination nodes / n5, n6 / ;
Parameter
   bw(f) Transmission rate (Kbps)
        / f1 32 / ;
Table delay(i,j) Link Delay
         n1       n2       n3       n4       n5       n6
```

```
n1        999       2         2         5         20        20
n2        999       999       999       999       2         999
n3        999       999       999       999       999       2
n4        999       999       999       999       5         5
n5        999       999       999       999       999       999
n6        999       999       999       999       999       999   ;
Parameter d(i,j) delay ;
   d(i,j)  =  delay(i,j)  ;
Table cap(i,j) Link Capacity (Kbps)
          n1        n2        n3        n4        n5        n6
n1        0         64        64        64        64        64
n2        0         0         0         0         64        0
n3        0         0         0         0         0         64
n4        0         0         0         0         64        64
n5        0         0         0         0         0         0
n6        0         0         0         0         0         0   ;
Parameter c(i,j) Link Capacity ;
   c(i,j)  =  cap(i,j)  ;
Table cost(i,j) Link Cost
          n1        n2        n3        n4        n5        n6
n1        999       2         2         5         20        20
n2        999       999       999       999       2         999
n3        999       999       999       999       999       2
n4        999       999       999       999       5         5
n5        999       999       999       999       999       999
n6        999       999       999       999       999       999   ;
Parameter w(i,j) delay ;
   w(i,j)  =  cost(i,j)  ;
Variables
   x(i,j,t,f)  If the link i j is used (0) to the flow f
with destination node t
   HC               Hop Count  (Hops)
   DL               Delay (ms)
   CT               Cost ($)
   BC               Bandwidth Consumption (Kbps)
   r1               r1
   r2               r2
   z                minimization ;
*Binary Variable x;
Positive Variable x;
```

```
*Positive Variable r1;
*Positive Variable r2;
Equations
Shortest_Path              Objective function
source_node(t,f)           source node
destination_node_5(f)      destination node
destination_node_6(f)      destination node
intermediate1_node(t,f)    intermediate node
intermediate2_node(t,f)      intermediate node
intermediate3_node(t,f)      intermediate node
*Link_capacity(i,j)        Link Capacity ;
*c1                          c1
*c2                          c2
*rs                          rs
;

*Shortest_Path        .. z =e= (r1*(sum((f), (sum((t),
sum((i,j), x(i,j,t,f)))))))
*                                   +
*                           (r2*(sum((f), (sum((t),
sum((i,j), d(i,j)*x(i,j,t,f)))))))   ;


Shortest_Path .. z =e= (1*(sum((f), (sum((t), sum((i,j),
x(i,j,t,f)))))))
                           +
                        (0*(sum((f), (sum((t), sum((i,j),
d(i,j)*x(i,j,t,f)))))))   ;

source_node(t,f)        .. sum((j), x('n1',j,t,f)) =e= 1;

destination_node_5(f) .. sum((i),x(i,'n5','n5',f)) =e= 1;
destination_node_6(f) .. sum((i),x(i,'n6','n6',f)) =e= 1;

intermediate1_node(t,f) ..sum((i,j), x(i,'n2',t,f)) -
sum((i,j), x('n2',j,t,f)) =e= 0;
intermediate2_node(t,f) .. sum((i,j), x(i,'n3',t,f)) -
sum((i,j), x('n3',j,t,f)) =e= 0;
intermediate3_node(t,f) .. sum((i,j), x(i,'n4',t,f)) -
sum((i,j), x('n4',j,t,f)) =e= 0;
```

```
*Link_Capacity(i,j) .. sum((f),bw(f)*smax((t),x(i,j,t,f)))
=l= c(i,j);


*c1              .. (sum((f), (sum((t), sum((i,j),
x(i,j,t,f))))))  =e= 3;
*c2              .. (sum((f), (sum((t), sum((i,j),
d(i,j)*x(i,j,t,f))))))  =e= 24;


*rs             .. r1+r2 =e= 1;



Model Transport /all/ ;

*option dnlp=SNOPT
option mip=CPLEX
*option iterlim = 50000
*option profiletol = 1.0E-2
solve transport using mip minimizing z ;
*solve transport using dnlp minimizing z ;

DL.l = sum((f), sum((t), sum((i,j), d(i,j)*x.l(i,j,t,f))));
HC.l = sum((f), sum((t), sum((i,j), x.l(i,j,t,f))));
CT.l = sum((f), sum((t), sum((i,j), w(i,j)*x.l(i,j,t,f))));
*BC.l = sum((f),sum((i,j),bw(f)*smax((t),x.l(i,j,t,f))));

Display x.l
Display DL.l
Display HC.l
Display CT.l
*Display r1.l
*Display r2.l
*Display BC.l
Display z.l
```

File in GAMS of Section 4.5.2.2 (f_1)

```
*****************************************************
***                                             ***
***          Multi-Objective Ej01               ***
***                                             ***
***       Objective Functions:                  ***
***       Hop Count                             ***
***       Delay                                 ***
***                                             ***
***       Eng. Yezid E. Donoso Meisel, Ph.D.    ***
*****************************************************
Sets
   i    ingress node / n1, n2, n3, n4, n5, n6 /
   j    egress node  / n1, n2, n3, n4, n5, n6 /
   f    Flows / f1 /
   t    destination nodes / n5, n6 / ;
Parameter
   bw(f) Transmission rate (Kbps)
         / f1 32 / ;
Table delay(i,j) Link Delay
          n1       n2       n3       n4       n5       n6
n1        999      2        2        5        20       20
n2        999      999      999      999      2        999
n3        999      999      999      999      999      2
n4        999      999      999      999      5        5
n5        999      999      999      999      999      999
n6        999      999      999      999      999      999    ;
Parameter d(i,j) delay ;
   d(i,j) = delay(i,j) ;
Table cap(i,j) Link Capacity (Kbps)
          n1       n2       n3       n4       n5       n6
n1        0        64       64       64       64       64
n2        0        0        0        0        64       0
n3        0        0        0        0        0        64
n4        0        0        0        0        64       64
n5        0        0        0        0        0        0
n6        0        0        0        0        0        0      ;
```

```
Parameter c(i,j) Link Capacity ;
  c(i,j) = cap(i,j) ;
Table cost(i,j) Link Cost
           n1       n2       n3       n4       n5       n6
n1         999      2        2        5        20       20
n2         999      999      999      999      2        999
n3         999      999      999      999      999      2
n4         999      999      999      999      5        5
n5         999      999      999      999      999      999
n6         999      999      999      999      999      999    ;
Parameter w(i,j) delay ;
  w(i,j) = cost(i,j) ;
Variables
  x(i,j,t,f)   If the link i j is used (0) to the flow f
with destination node t
  HC              Hop Count (Hops)
  DL              Delay (ms)
  CT              Cost ($)
  BC              Bandwidth Consumption (Kbps)
  r1              r1
  r2              r2
  z               minimization ;
Binary Variable x;
*Positive Variable x;
*Positive Variable r1;
*Positive Variable r2;
Equations
Shortest_Path              Objective function
e_constraint_f1            constraint function f2
source_node(t,f)           source node
destination_node_5(f)      destination node
destination_node_6(f)      destination node
intermediate1_node(t,f)     intermediate node
intermediate2_node(t,f)     intermediate node
intermediate3_node(t,f)     intermediate node
*Link_capacity(i,j)        Link Capacity ;
*c1                         c1
*c2                         c2
*rs                         rs
;
```

```
*Shortest_Path            .. z =e= (r1*(sum((f), (sum((t),
sum((i,j), x(i,j,t,f)))))))
*                                   +
*                                   (r2*(sum((f), (sum((t),
sum((i,j), d(i,j)*x(i,j,t,f)))))))   ;

Shortest_Path .. z =e= sum((f), (sum((t), sum((i,j),
d(i,j)*x(i,j,t,f)))))   ;

e_constraint_f1          .. (sum((f), (sum((t), sum((i,j),
x(i,j,t,f)))))) =l= 2;

source_node(t,f)         .. sum((j), x('n1',j,t,f)) =e= 1;

destination_node_5(f) .. sum((i),x(i,'n5','n5',f)) =e= 1;
destination_node_6(f) .. sum((i),x(i,'n6','n6',f)) =e= 1;

intermediate1_node(t,f).. sum((i,j), x(i,'n2',t,f)) -
sum((i,j), x('n2',j,t,f)) =e= 0;
intermediate2_node(t,f) .. sum((i,j), x(i,'n3',t,f)) -
sum((i,j), x('n3',j,t,f)) =e= 0;
intermediate3_node(t,f) .. sum((i,j), x(i,'n4',t,f)) -
sum((i,j), x('n4',j,t,f)) =e= 0;

*Link_Capacity(i,j) .. sum((f),bw(f)*smax((t),x(i,j,t,f)))
=l= c(i,j);

*c1              .. (sum((f), (sum((t), sum((i,j),
x(i,j,t,f)))))) =e= 3;
*c2              .. (sum((f), (sum((t), sum((i,j),
d(i,j)*x(i,j,t,f)))))) =e= 24;

*rs              .. r1+r2 =e= 1;


Model Transport /all/ ;

*option dnlp=SNOPT
option mip=CPLEX
*option iterlim = 50000
*option profiletol = 1.0E-2
```

```
solve transport using mip minimizing z ;
*solve transport using dnlp minimizing z ;

DL.l = sum((f), sum((t), sum((i,j), d(i,j)*x.l(i,j,t,f))));
HC.l = sum((f), sum((t), sum((i,j), x.l(i,j,t,f))));
CT.l = sum((f), sum((t), sum((i,j), w(i,j)*x.l(i,j,t,f))));
*BC.l = sum((f), sum((i,j),
bw(f)*smax((t),x.l(i,j,t,f))));

Display x.l
Display DL.l
Display HC.l
Display CT.l
*Display r1.l
*Display r2.l
*Display BC.l
Display z.l
```

File in GAMS of Section 4.5.2.2 (f_2)

```
* * * * * * * * * * * * * * * * * * * * * * * * * * * * * * * * * * * * * * * * * * * *
***                                                          ***
***            Multi-Objective Ej01                          ***
***    ***
***          Objective Functions:                            ***
***          Hop Count                                       ***
***          Delay                                           ***
***                                                          ***
***       Eng. Yezid E. Donoso Meisel, Ph.D.                 ***
* * * * * * * * * * * * * * * * * * * * * * * * * * * * * * * * * * * * * * * * * * * *
Sets
   i    ingress node / n1, n2, n3, n4, n5, n6 /
   j    egress node  / n1, n2, n3, n4, n5, n6 /
   f    Flows / f1 /
   t    destination nodes / n5, n6 / ;
Parameter
   bw(f) Transmission rate (Kbps)
         / f1 32 / ;
```

```
Table delay(i,j)  Link Delay
          n1        n2        n3        n4        n5        n6
n1        999       2         2         5         20        20
n2        999       999       999       999       2         999
n3        999       999       999       999       999       2
n4        999       999       999       999       5         5
n5        999       999       999       999       999       999
n6        999       999       999       999       999       999   ;
Parameter d(i,j) delay ;
   d(i,j) = delay(i,j) ;
Table cap(i,j) Link Capacity (Kbps)
          n1        n2        n3        n4        n5        n6
n1        0         64        64        64        64        64
n2        0         0         0         0         64        0
n3        0         0         0         0         0         64
n4        0         0         0         0         64        64
n5        0         0         0         0         0         0
n6        0         0         0         0         0         0   ;
Parameter c(i,j) Link Capacity ;
   c(i,j) = cap(i,j) ;
Table cost(i,j) Link Cost
          n1        n2        n3        n4        n5        n6
n1        999       2         2         5         20        20
n2        999       999       999       999       2         999
n3        999       999       999       999       999       2
n4        999       999       999       999       5         5
n5        999       999       999       999       999       999
n6        999       999       999       999       999       999   ;
Parameter w(i,j) delay ;
   w(i,j) = cost(i,j) ;
Variables
   x(i,j,t,f)  If the link i j is used (0) to the flow f
with destination node t
   HC             Hop Count (Hops)
   DL             Delay (ms)
   CT             Cost ($)
   BC             Bandwidth Consumption (Kbps)
   r1             r1
   r2             r2
   z              minimization ;
```

```
Binary Variable x;
*Positive Variable x;
*Positive Variable r1;
*Positive Variable r2;
Equations
Shortest_Path              Objective function
e_constraint_f2            constraint function f2
source_node(t,f)           source node
destination_node_5(f)      destination node
destination_node_6(f)      destination node
intermediate1_node(t,f)     intermediate node
intermediate2_node(t,f)     intermediate node
intermediate3_node(t,f)     intermediate node
*Link_capacity(i,j)        Link Capacity ;
*c1                         c1
*c2                         c2
*rs                         rs
;


*Shortest_Path         .. z =e= (r1*(sum((f), (sum((t),
sum((i,j), x(i,j,t,f)))))))
*                                +
*                                (r2*(sum((f), (sum((t),
sum((i,j), d(i,j)*x(i,j,t,f)))))))   ;


Shortest_Path .. z =e= (sum((f), (sum((t), sum((i,j),
x(i,j,t,f))))))   ;

e_constraint_f2        .. sum((f), (sum((t), sum((i,j),
d(i,j)*x(i,j,t,f))))) =l= 10;

source_node(t,f)       .. sum((j), x('n1',j,t,f)) =e= 1;

destination_node_5(f) .. sum((i),x(i,'n5','n5',f)) =e= 1;
destination_node_6(f) .. sum((i),x(i,'n6','n6',f)) =e= 1;

intermediate1_node(t,f) .. sum((i,j), x(i,'n2',t,f)) -
sum((i,j), x('n2',j,t,f)) =e= 0;
intermediate2_node(t,f) .. sum((i,j), x(i,'n3',t,f)) -
sum((i,j), x('n3',j,t,f)) =e= 0;
```

```
intermediate3_node(t,f)  .. sum((i,j),  x(i,'n4',t,f)) -
sum((i,j),  x('n4',j,t,f))  =e=  0;


*Link_Capacity(i,j) .. sum((f),bw(f)*smax((t),x(i,j,t,f)))
=l=  c(i,j);


*c1            .. (sum((f),  (sum((t),  sum((i,j),
x(i,j,t,f))))))  =e=  3;
*c2            .. (sum((f),  (sum((t),  sum((i,j),
d(i,j)*x(i,j,t,f))))))  =e=  24;


*rs            ..  r1+r2 =e=  1;



Model Transport /all/ ;


*option dnlp=SNOPT
option mip=CPLEX
*option iterlim = 50000
*option profiletol = 1.0E-2
solve transport using mip minimizing z ;
*solve transport using dnlp minimizing z ;


DL.l = sum((f), sum((t), sum((i,j), d(i,j)*x.l(i,j,t,f))));
HC.l = sum((f), sum((t), sum((i,j), x.l(i,j,t,f))));
CT.l = sum((f), sum((t), sum((i,j), w(i,j)*x.l(i,j,t,f))));
*BC.l = sum((f), sum((i,j),
bw(f)*smax((t),x.l(i,j,t,f))));


Display x.l
Display DL.l
Display HC.l
Display CT.l
*Display r1.l
*Display r2.l
*Display BC.l
Display z.l
```

File in GAMS of Section 4.5.2.3 (*r* = 2)

```
*************************************************
***                                         ***
***        Multi-Objective Ej01             ***
***                                         ***
***        Objective Functions:             ***
***        Hop Count                        ***
***        Delay                            ***
***                                         ***
***     Eng. Yezid E. Donoso Meisel, Ph.D.  ***
*************************************************
Sets
   i   ingress node / n1, n2, n3, n4, n5, n6 /
   j   egress node  / n1, n2, n3, n4, n5, n6 /
   f   Flows / f1 /
   t   destination nodes / n5, n6 / ;
Parameter
   bw(f) Transmission rate (Kbps)
        / f1 32 / ;
Table delay(i,j) Link Delay
            n1      n2      n3      n4      n5      n6
n1          999     2       2       5       20      20
n2          999     999     999     999     2       999
n3          999     999     999     999     999     2
n4          999     999     999     999     5       5
n5          999     999     999     999     999     999
n6          999     999     999     999     999     999  ;
Parameter d(i,j) delay ;
   d(i,j) = delay(i,j) ;
Table cap(i,j) Link Capacity (Kbps)
            n1      n2      n3      n4      n5      n6
n1          0       64      64      64      64      64
n2          0       0       0       0       64      0
n3          0       0       0       0       0       64
n4          0       0       0       0       64      64
n5          0       0       0       0       0       0
n6          0       0       0       0       0       0   ;
Parameter c(i,j) Link Capacity ;
```

```
   c(i,j) = cap(i,j) ;
Table cost(i,j) Link Cost
          n1       n2       n3       n4       n5       n6
n1        999      2        2        5        20       20
n2        999      999      999      999      2        999
n3        999      999      999      999      999      2
n4        999      999      999      999      5        5
n5        999      999      999      999      999      999
n6        999      999      999      999      999      999   ;
Parameter w(i,j) delay ;
   w(i,j) = cost(i,j) ;
Variables
   x(i,j,t,f)  If the link i j is used (0) to the flow f
with destination node t
   HC             Hop Count (Hops)
   DL             Delay (ms)
   CT             Cost ($)
   BC             Bandwidth Consumption (Kbps)
   r1             r1
   r2             r2
   z              minimization ;
Binary Variable x;
*Positive Variable x;
*Positive Variable r1;
*Positive Variable r2;
Equations
Shortest_Path         Objective function
source_node(t,f)              source node
destination_node_5(f)           destination node
destination_node_6(f)           destination node
intermediate1_node(t,f)     intermediate node
intermediate2_node(t,f)     intermediate node
intermediate3_node(t,f)     intermediate node
*Link_capacity(i,j)   Link Capacity ;
*c1                   c1
*c2                   c2
*rs                   rs
;
```

```
Shortest_Path .. z =e= sqrt((0.1*sqr(abs(5-(sum((f),
(sum((t), sum((i,j), x(i,j,t,f)))))))))
                            +
                               (0.9*sqr(abs(10-(sum((f),
(sum((t), sum((i,j), d(i,j)*x(i,j,t,f)))))))))))   ;




source_node(t,f)        .. sum((j), x('n1',j,t,f)) =e= 1;

destination_node_5(f) .. sum((i),x(i,'n5','n5',f)) =e= 1;
destination_node_6(f) .. sum((i),x(i,'n6','n6',f)) =e= 1;

intermediate1_node(t,f) .. sum((i,j), x(i,'n2',t,f)) -
sum((i,j), x('n2',j,t,f)) =e= 0;
intermediate2_node(t,f) .. sum((i,j), x(i,'n3',t,f)) -
sum((i,j), x('n3',j,t,f)) =e= 0;
intermediate3_node(t,f) .. sum((i,j), x(i,'n4',t,f)) -
sum((i,j), x('n4',j,t,f)) =e= 0;

*Link_Capacity(i,j) .. sum((f),bw(f)*smax((t),x(i,j,t,f)))
=l= c(i,j);

*c1             .. (sum((f), (sum((t), sum((i,j),
x(i,j,t,f)))))) =e= 3;
*c2             .. (sum((f), (sum((t), sum((i,j),
d(i,j)*x(i,j,t,f)))))) =e= 24;

*rs             .. r1+r2 =e= 1;



Model Transport /all/ ;

*option dnlp=SNOPT
*option mip=CPLEX
option minlp=SBB
*option iterlim = 50000
*option profiletol = 1.0E-2
solve transport using minlp minimizing z ;
*solve transport using dnlp minimizing z ;
```

```
DL.l = sum((f), sum((t), sum((i,j), d(i,j)*x.l(i,j,t,f))));
HC.l = sum((f), sum((t), sum((i,j), x.l(i,j,t,f))));
CT.l = sum((f), sum((t), sum((i,j), w(i,j)*x.l(i,j,t,f))));
*BC.l = sum((f), sum((i,j),
bw(f)*smax((t),x.l(i,j,t,f))));

Display x.l
Display DL.l
Display HC.l
Display CT.l
*Display r1.l
*Display r2.l
*Display BC.l
Display z.l
```

File in GAMS of Section 4.5.2.3 ($r = \infty$)

```
* * * * * * * * * * * * * * * * * * * * * * * * * * * * * * * * * * * * * * * *
***                                                          ***
***          Multi-Objective Ej01                           ***
***                                                          ***
***          Objective Functions:                           ***
***          Hop Count                                       ***
***          Delay                                           ***
***                                                          ***
***          Eng. Yezid E. Donoso Meisel, Ph.D.             ***
* * * * * * * * * * * * * * * * * * * * * * * * * * * * * * * * * * * * * * * *
Sets
   i    ingress node / n1, n2, n3, n4, n5, n6 /
   j    egress node  / n1, n2, n3, n4, n5, n6 /
   f    Flows / f1 /
   t    destination nodes / n5, n6 / ;
Parameter
   bw(f) Transmission rate (Kbps)
        / f1 32 / ;
Table delay(i,j) Link Delay
```

	n1	n2	n3	n4	n5	n6
n1	999	2	2	5	20	20
n2	999	999	999	999	2	999
n3	999	999	999	999	999	2
n4	999	999	999	999	5	5
n5	999	999	999	999	999	999
n6	999	999	999	999	999	999 ;

```
Parameter d(i,j) delay ;
  d(i,j) = delay(i,j) ;
Table cap(i,j) Link Capacity (Kbps)
```

	n1	n2	n3	n4	n5	n6
n1	0	64	64	64	64	64
n2	0	0	0	0	64	0
n3	0	0	0	0	0	64
n4	0	0	0	0	64	64
n5	0	0	0	0	0	0
n6	0	0	0	0	0	0 ;

```
Parameter c(i,j) Link Capacity ;
  c(i,j) = cap(i,j) ;
Table cost(i,j) Link Cost
```

	n1	n2	n3	n4	n5	n6
n1	999	2	2	5	20	20
n2	999	999	999	999	2	999
n3	999	999	999	999	999	2
n4	999	999	999	999	5	5
n5	999	999	999	999	999	999
n6	999	999	999	999	999	999 ;

```
Parameter w(i,j) delay ;
  w(i,j) = cost(i,j) ;
Variables
  x(i,j,t,f)  If the link i j is used (0) to the flow f
with destination node t
  HC          Hop Count (Hops)
  DL          Delay (ms)
  CT          Cost ($)
  BC          Bandwidth Consumption (Kbps)
  r1          r1
  r2          r2
  z           minimization ;
Binary Variable x;
```

```
*Positive Variable x;
*Positive Variable r1;
*Positive Variable r2;
Equations
Shortest_Path              Objective function
source_node(t,f)           source node
destination_node_5(f)      destination node
destination_node_6(f)      destination node
intermediate1_node(t,f)      intermediate node
intermediate2_node(t,f)      intermediate node
intermediate3_node(t,f)      intermediate node
*Link_capacity(i,j)        Link Capacity ;
*c1                          c1
*c2                          c2
*rs                          rs
;



Shortest_Path          .. z =e= max((0.1*abs(3-(sum((f),
(sum((t), sum((i,j), x(i,j,t,f)))))))),
                           (0.9*abs(24-(sum((f), (sum((t),
sum((i,j), d(i,j)*x(i,j,t,f))))))))))   ;



source_node(t,f)        .. sum((j), x('n1',j,t,f)) =e= 1;

destination_node_5(f) .. sum((i),x(i,'n5','n5',f)) =e= 1;
destination_node_6(f) .. sum((i),x(i,'n6','n6',f)) =e= 1;

intermediate1_node(t,f) .. sum((i,j), x(i,'n2',t,f)) -
sum((i,j), x('n2',j,t,f)) =e= 0;
intermediate2_node(t,f) .. sum((i,j), x(i,'n3',t,f)) -
sum((i,j), x('n3',j,t,f)) =e= 0;
intermediate3_node(t,f) .. sum((i,j), x(i,'n4',t,f)) -
sum((i,j), x('n4',j,t,f)) =e= 0;

*Link_Capacity(i,j) .. sum((f),bw(f)*smax((t),x(i,j,t,f)))
=l= c(i,j);
```

```
*c1                    .. (sum((f), (sum((t), sum((i,j),
x(i,j,t,f)))))) =e= 3;
*c2                    .. (sum((f), (sum((t), sum((i,j),
d(i,j)*x(i,j,t,f)))))) =e= 24;


*rs                    .. r1+r2 =e= 1;



Model Transport /all/ ;

*option dnlp=SNOPT
*option mip=CPLEX
option minlp=SBB
*option iterlim = 50000
*option profiletol = 1.0E-2
solve transport using minlp minimizing z ;
*solve transport using dnlp minimizing z ;

DL.l = sum((f), sum((t), sum((i,j), d(i,j)*x.l(i,j,t,f))));
HC.l = sum((f), sum((t), sum((i,j), x.l(i,j,t,f))));
CT.l = sum((f), sum((t), sum((i,j), w(i,j)*x.l(i,j,t,f))));
*BC.l = sum((f), sum((i,j),
bw(f)*smax((t),x.l(i,j,t,f))));

Display x.l
Display DL.l
Display HC.l
Display CT.l
*Display r1.l
*Display r2.l
*Display BC.l
Display z.l
```

File in GAMS of Section 4.5.2.4

```
***************************************************
***                                             ***
***          Multi-Objective Ej01               ***
***                                             ***
***       Objective  Functions:                 ***
***       Hop Count                             ***
***       Delay                                 ***
***                                             ***
***       Eng. Yezid E. Donoso Meisel, Ph.D.    ***
***************************************************
Sets
   i    ingress node / n1, n2, n3, n4, n5, n6 /
   j    egress node  / n1, n2, n3, n4, n5, n6 /
   f    Flows / f1 /
   t    destination nodes / n5, n6 / ;
Parameter
   bw(f)  Transmission rate (Kbps)
        / f1 32 / ;
Table delay(i,j) Link Delay
          n1      n2      n3      n4      n5      n6
n1        999     2       2       5       20      20
n2        999     999     999     999     2       999
n3        999     999     999     999     999     2
n4        999     999     999     999     5       5
n5        999     999     999     999     999     999
n6        999     999     999     999     999     999  ;
Parameter d(i,j) delay ;
   d(i,j) = delay(i,j)  ;
Table cap(i,j) Link Capacity (Kbps)
          n1      n2      n3      n4      n5      n6
n1        0       64      64      64      64      64
n2        0       0       0       0       64      0
n3        0       0       0       0       0       64
n4        0       0       0       0       64      64
n5        0       0       0       0       0       0
n6        0       0       0       0       0       0  ;
```

```
Parameter c(i,j) Link Capacity ;
  c(i,j) = cap(i,j) ;
Table cost(i,j) Link Cost
           n1        n2        n3        n4        n5        n6
n1         999       2         2         5         20        20
n2         999       999       999       999       2         999
n3         999       999       999       999       999       2
n4         999       999       999       999       5         5
n5         999       999       999       999       999       999
n6         999       999       999       999       999       999   ;
Parameter w(i,j) delay ;
  w(i,j) = cost(i,j) ;
Variables
  x(i,j,t,f)  If the link i j is used (0) to the flow f
with destination node t
  HC              Hop Count (Hops)
  DL              Delay (ms)
  CT              Cost ($)
  BC              Bandwidth Consumption (Kbps)
  r1              r1
  r2              r2
  z               minimization ;
Binary Variable x;
*Positive Variable x;
*Positive Variable r1;
*Positive Variable r2;
Equations
Shortest_Path           Objective function
e_constraint_f1           constraint function f1
e_constraint_f2           constraint function f2
source_node(t,f)            source node
destination_node_5(f)         destination node
destination_node_6(f)         destination node
intermediate1_node(t,f)     intermediate node
intermediate2_node(t,f)     intermediate node
intermediate3_node(t,f)     intermediate node
*Link_capacity(i,j)    Link Capacity ;
*c1                     c1
*c2                     c2
*rs                     rs
;
```

```
Shortest_Path .. z =e= max(0,(3-(sum((f), (sum((t),
sum((i,j), x(i,j,t,f)))))))
                   +
                        max(0,(24-(sum((f), (sum((t),
sum((i,j), d(i,j)*x(i,j,t,f)))))))))   ;



e_constraint_f1       .. sum((f), (sum((t), sum((i,j),
x(i,j,t,f))))) =l= 5;
e_constraint_f2       .. sum((f), (sum((t), sum((i,j),
d(i,j)*x(i,j,t,f))))) =l= 25;




source_node(t,f)      .. sum((j), x('n1',j,t,f)) =e= 1;

destination_node_5(f) .. sum((i),x(i,'n5','n5',f)) =e= 1;
destination_node_6(f) .. sum((i),x(i,'n6','n6',f)) =e= 1;

intermediate1_node(t,f) .. sum((i,j), x(i,'n2',t,f)) -
sum((i,j), x('n2',j,t,f)) =e= 0;
intermediate2_node(t,f) .. sum((i,j), x(i,'n3',t,f)) -
sum((i,j), x('n3',j,t,f)) =e= 0;
intermediate3_node(t,f) .. sum((i,j), x(i,'n4',t,f)) -
sum((i,j), x('n4',j,t,f)) =e= 0;

*Link_Capacity(i,j) .. sum((f),bw(f)*smax((t),x(i,j,t,f)))
=l= c(i,j);

*c1                   .. (sum((f), (sum((t), sum((i,j),
x(i,j,t,f)))))) =e= 3;
*c2                   .. (sum((f), (sum((t), sum((i,j),
d(i,j)*x(i,j,t,f)))))) =e= 24;


*rs                   .. r1+r2 =e= 1;



Model Transport /all/ ;

*option dnlp=SNOPT
*option mip=CPLEX
```

```
option minlp=SBB
*option iterlim = 50000
*option profiletol = 1.0E-2
solve transport using minlp maximizing z ;
*solve transport using dnlp minimizing z ;

DL.l = sum((f), sum((t), sum((i,j), d(i,j)*x.l(i,j,t,f))));
HC.l = sum((f), sum((t), sum((i,j), x.l(i,j,t,f))));
CT.l = sum((f), sum((t), sum((i,j), w(i,j)*x.l(i,j,t,f))));
*BC.l = sum((f), sum((i,j),
bw(f)*smax((t),x.l(i,j,t,f))));

Display x.l
Display DL.l
Display HC.l
Display CT.l
*Display r1.l
*Display r2.l
*Display BC.l
Display z.l
```

File in GAMS of Section 4.5.3 (Examples 1 and 2)

```
*********************************************************
***                                                 ***
***         Multi-Objective  NSF  Network           ***
***                                                 ***
***       Objective  Functions:                      ***
***       Hop  Count                                 ***
***       Delay                                       ***
***       Bandwidth  Consumption                      ***
***                                                 ***
***       Eng.  Yezid  E.  Donoso  Meisel,  Ph.D.    ***
*********************************************************
Sets
   i    ingress node / n0, n1, n2, n3, n4, n5, n6, n7, n8,
n9, n10, n11, n12, n13 /
```

```
    j    egress node   / n0, n1, n2, n3, n4, n5, n6, n7,
n8, n9, n10, n11, n12, n13 /
    f    Flows / f1 / ;
Parameter
   bw(f) Transmission rate (Kbps)
         / f1 512 / ;
Table delay(i,j) Link Delay
```

	n0	n1	n2	n3	n4	n5	n6	n7	n8	n9	n10	n11	n12	n13
n0	999	9	9	7	999	999	999	999	999	999	999	999	999	999
n1	999	999	999	13	999	999	20	999	999	999	999	999	999	999
n2	999	999	999	999	7	999	999	26	999	999	999	999	999	999
n3	999	13	999	999	999	999	999	999	999	999	15	999	999	999
n4	999	999	7	999	999	7	999	999	999	999	11	999	999	999
n5	999	999	999	999	7	999	7	999	999	999	999	999	999	999
n6	999	20	999	999	999	7	999	999	999	7	999	999	999	999
n7	999	999	26	999	999	999	999	999	5	999	999	999	999	8
n8	999	999	999	999	999	999	999	5	999	5	999	999	7	999
n9	999	999	999	999	999	999	7	999	5	999	999	8	999	999
n10	999	999	999	15	11	999	999	999	999	999	999	9	14	999
n11	999	999	999	999	999	999	999	999	999	8	9	999	999	999
n12	999	999	999	999	999	999	999	999	7	999	14	999	999	4
n13	999	999	999	999	999	999	999	8	999	999	999	999	4	999 ;

```
   Parameter d(i,j) delay ;
     d(i,j) = delay(i,j)  ;
   Table cap(i,j) Link Capacity (Kbps)
```

	n0	n1	n2	n3	n4	n5	n6	n7	n8	n9	n10	n11	n12	n13
n0	0	1536	1536	1536	0	0	0	0	0	0	0	0	0	0
n1	0	0	0	1536	0	0	1536	0	0	0	0	0	0	0
n2	0	0	0	0	1536	0	0	1536	0	0	0	0	0	0
n3	0	1536	0	0	0	0	0	0	0	0	1536	0	0	0
n4	0	0	1536	0	0	1536	0	0	0	0	1536	0	0	0
n5	0	0	0	0	1536	0	1536	0	0	0	0	0	0	0
n6	0	1536	0	0	0	1536	0	0	0	1536	0	0	0	0
n7	0	0	1536	0	0	0	0	0	1536	0	0	0	0	1536
n8	0	0	0	0	0	0	0	1536	0	1536	0	0	1536	0
n9	0	0	0	0	0	0	1536	0	1536	0	0	1536	0	0
n10	0	0	0	1536	1536	0	0	0	0	0	0	1536	1536	0
n11	0	0	0	0	0	0	0	0	0	1536	1536	0	0	0
n12	0	0	0	0	0	0	0	0	1536	0	1536	0	0	1536
n13	0	0	0	0	0	0	0	1536	0	0	0	0	1536	0 ;

```
Parameter c(i,j) Link Capacity ;
```

```
   c(i,j)  =  cap(i,j)  ;
Table cost(i,j) Link Cost
```

n0	n1	n2	n3	n4	n5	n6	n7	n8	n9	n10	n11	n12	n13
n0 999	9	9	7	999	999	999	999	999	999	999	999	999	999
n1 999	999	999	13	999	999	20	999	999	999	999	999	999	999
n2 999	999	999	999	7	999	999	26	999	999	999	999	999	999
n3 999	13	999	999	999	999	999	999	999	999	15	999	999	999
n4 999	999	7	999	999	7	999	999	999	999	11	999	999	999
n5 999	999	999	999	7	999	7	999	999	999	999	999	999	999
n6 999	20	999	999	999	7	999	999	999	7	999	999	999	999
n7 999	999	26	999	999	999	999	999	5	999	999	999	999	8
n8 999	999	999	999	999	999	999	5	999	5	999	999	7	999
n9 999	999	999	999	999	999	7	999	5	999	999	8	999	999
n10 999	999	999	15	11	999	999	999	999	999	999	9	14	999
n11 999	999	999	999	999	999	999	999	999	8	9	999	999	999
n12 999	999	999	999	999	999	999	999	7	999	14	999	999	4
n13 999	999	999	999	999	999	999	8	999	999	999	999	4	999 ;

```
Parameter w(i,j) cost ;
   w(i,j) = cost(i,j) ;
Variables
   x(i,j,f) If the link i j is used (0) to the flow f
   HC      Hop Count (Hops)
   DL      Delay (ms)
   CT      Cost ($)
   BC      Bandwidth Consumption (Kbps)
   z       minimization ;
Binary Variable x;
Equations
Shortest_Path            Objective function
source_node(f)              source node
destination_node(f)           destination node
intermediate1_node(f)      intermediate node
intermediate2_node(f)      intermediate node
intermediate3_node(f)      intermediate node
intermediate4_node(f)      intermediate node
intermediate5_node(f)      intermediate node
intermediate6_node(f)      intermediate node
intermediate7_node(f)      intermediate node
intermediate8_node(f)      intermediate node
intermediate9_node(f)      intermediate node
```

```
intermediate10_node(f)      intermediate node
intermediate11_node(f)        intermediate node
intermediate12_node(f)        intermediate node
Link_capacity(i,j)      Link Capacity ;


Shortest_Path                   .. z =e=  (0.3*(sum((f),sum((i,j),
x(i,j,f)))))

                                    +
                                    (0.3*(sum((f),sum((i,j),
d(i,j)*x(i,j,f)))))

                                    +
                                    (0.4*(sum((f),sum((i,j),
bw(f)*x(i,j,f)))))  ;


source_node(f)                  .. sum((j), x('n0',j,f)) =e= 1;


destination_node(f)             .. sum((i),x(i,'n13',f)) =e= 1;


intermediate1_node(f) .. sum((i,j), x(i,'n1',f)) -
sum((i,j), x('n1',j,f)) =e= 0;
intermediate2_node(f) .. sum((i,j), x(i,'n2',f)) -
sum((i,j), x('n2',j,f)) =e= 0;
intermediate3_node(f) .. sum((i,j), x(i,'n3',f)) -
sum((i,j), x('n3',j,f)) =e= 0;
intermediate4_node(f) .. sum((i,j), x(i,'n4',f)) -
sum((i,j), x('n4',j,f)) =e= 0;
intermediate5_node(f) .. sum((i,j), x(i,'n5',f)) -
sum((i,j), x('n5',j,f)) =e= 0;
intermediate6_node(f) .. sum((i,j), x(i,'n6',f)) -
sum((i,j), x('n6',j,f)) =e= 0;
intermediate7_node(f) .. sum((i,j), x(i,'n7',f)) -
sum((i,j), x('n7',j,f)) =e= 0;
intermediate8_node(f) .. sum((i,j), x(i,'n8',f)) -
sum((i,j), x('n8',j,f)) =e= 0;
intermediate9_node(f) .. sum((i,j), x(i,'n9',f)) -
sum((i,j), x('n9',j,f)) =e= 0;
intermediate10_node(f).. sum((i,j), x(i,'n10',f)) -
sum((i,j), x('n10',j,f)) =e= 0;
intermediate11_node(f).. sum((i,j), x(i,'n11',f)) -
sum((i,j), x('n11',j,f)) =e= 0;
intermediate12_node(f).. sum((i,j), x(i,'n12',f)) -
sum((i,j), x('n12',j,f)) =e= 0;
```

```
Link_Capacity(i,j)      .. sum((f),bw(f)*x(i,j,f)) =l=
c(i,j);

Model Transport /all/ ;

option mip=CPLEX
*option iterlim = 50000
*option profiletol = 1.0E-2
Solve transport using mip minimizing z ;

DL.l = sum((f),sum((i,j), d(i,j)*x.l(i,j,f)));
HC.l = sum((f),sum((i,j), x.l(i,j,f)));
CT.l = sum((f),sum((i,j), w(i,j)*x.l(i,j,f)));
BC.l = sum((f),sum((i,j), bw(f)*x.l(i,j,f)));

Display x.l
Display DL.l
Display HC.l
Display CT.l
Display BC.l
Display z.l
```

File in GAMS of Section 4.5.3 (Examples 3 and 4)

```
********************************************************
***                                                ***
***         Multi-Objective NSF Network            ***
***                                                ***
***       Objective Functions:                     ***
***       Hop Count                                ***
***       Delay                                    ***
***       Bandwidth Consumption                    ***
***                                                ***
***     Eng. Yezid E. Donoso Meisel, Ph.D.         ***
********************************************************
Sets
```

```
    i    ingress node / n0, n1, n2, n3, n4, n5, n6, n7, n8,
n9, n10, n11, n12, n13 /
    j    egress node  / n0, n1, n2, n3, n4, n5, n6, n7,
n8, n9, n10, n11, n12, n13 /
    f    Flows / f1, f2 / ;
Parameter
    bw(f) Transmission rate (Kbps)
          / f1 1024
            f2 1024 / ;
Table delay(i,j) Link Delay
```

	n0	n1	n2	n3	n4	n5	n6	n7	n8	n9	n10	n11	n12	n13
n0	999	9	9	7	999	999	999	999	999	999	999	999	999	999
n1	999	999	999	13	999	999	20	999	999	999	999	999	999	999
n2	999	999	999	999	7	999	999	26	999	999	999	999	999	999
n3	999	13	999	999	999	999	999	999	999	999	15	999	999	999
n4	999	999	7	999	999	7	999	999	999	999	11	999	999	999
n5	999	999	999	999	7	999	7	999	999	999	999	999	999	999
n6	999	20	999	999	999	7	999	999	999	7	999	999	999	999
n7	999	999	26	999	999	999	999	999	5	999	999	999	999	8
n8	999	999	999	999	999	999	999	5	999	5	999	999	7	999
n9	999	999	999	999	999	999	7	999	5	999	999	8	999	999
n10	999	999	999	15	11	999	999	999	999	999	999	9	14	999
n11	999	999	999	999	999	999	999	999	999	8	9	999	999	999
n12	999	999	999	999	999	999	999	999	7	999	14	999	999	4
n13	999	999	999	999	999	999	999	8	999	999	999	999	4	999;

```
Parameter d(i,j) delay ;
    d(i,j) = delay(i,j) ;
```

```
Table cap(i,j) Link Capacity (Kbps)
```

	n0	n1	n2	n3	n4	n5	n6	n7	n8	n9	n10	n11	n12	n13
n0	0	1536	1536	1536	0	0	0	0	0	0	0	0	0	0
n1	0	0	0	1536	0	0	1536	0	0	0	0	0	0	0
n2	0	0	0	0	1536	0	0	1536	0	0	0	0	0	0
n3	0	1536	0	0	0	0	0	0	0	0	1536	0	0	0
n4	0	0	1536	0	0	1536	0	0	0	0	1536	0	0	0
n5	0	0	0	0	1536	0	1536	0	0	0	0	0	0	0
n6	0	1536	0	0	0	1536	0	0	0	153	0	0	0	0
n7	0	0	1536	0	0	0	0	0	1536	0	0	0	0	1536
n8	0	0	0	0	0	0	0	1536	0	1536	0	0	1536	0
n9	0	0	0	0	0	0	1536	0	1536	0	0	1536	0	0
n10	0	0	0	1536	1536	0	0	0	0	0	0	1536	1536	0
n11	0	0	0	0	0	0	0	0	0	1536	1536	0	0	0
n12	0	0	0	0	0	0	0	0	1536	0	1536	0	0	1536
n13	0	0	0	0	0	0	0	1536	0	0	0	0	1536	0 ;

```
Parameter c(i,j) Link Capacity ;
   c(i,j)  =  cap(i,j)  ;
Table cost(i,j) Link Cost
```

	n0	n1	n2	n3	n4	n5	n6	n7	n8	n9	n10	n11	n12	n13
n0	999	9	9	7	999	999	999	999	999	999	999	999	999	999
n1	999	999	999	13	999	999	20	999	999	999	999	999	999	999
n2	999	999	999	999	7	999	999	26	999	999	999	999	999	999
n3	999	13	999	999	999	999	999	999	999	999	15	999	999	999
n4	999	999	7	999	999	7	999	999	999	999	11	999	999	999
n5	999	999	999	999	7	999	7	999	999	999	999	999	999	999
n6	999	20	999	999	999	7	999	999	999	7	999	999	999	999
n7	999	999	26	999	999	999	999	999	5	999	999	999	999	8
n8	999	999	999	999	999	999	999	5	999	5	999	999	7	999
n9	999	999	999	999	999	999	7	999	5	999	999	8	999	999
n10	999	999	999	15	11	999	999	999	999	999	999	9	14	999
n11	999	999	999	999	999	999	999	999	999	8	9	999	999	999
n12	999	999	999	999	999	999	999	999	7	999	14	999	999	4
n13	999	999	999	999	999	999	999	8	999	999	999	999	4	999 ;

```
Parameter w(i,j) cost ;
   w(i,j)  =  cost(i,j)  ;
Variables
   x(i,j,f)  If the link i j is used (0) to the flow f
   HC      Hop Count (Hops)
   DL      Delay (ms)
   CT      Cost ($)
```

```
   BC      Bandwidth Consumption (Kbps)
   z       minimization ;
Binary Variable x;
Equations
Shortest_Path              Objective function
source_node(f)                   source node
destination_node(f)                destination node
intermediate1_node(f)      intermediate node
intermediate2_node(f)      intermediate node
intermediate3_node(f)      intermediate node
intermediate4_node(f)      intermediate node
intermediate5_node(f)      intermediate node
intermediate6_node(f)      intermediate node
intermediate7_node(f)      intermediate node
intermediate8_node(f)      intermediate node
intermediate9_node(f)      intermediate node
intermediate10_node(f)      intermediate node
intermediate11_node(f)      intermediate node
intermediate11_node(f)      intermediate node
Link_capacity(i,j)    Link Capacity ;

Shortest_Path              .. z =e= (0.3*(sum((f),sum((i,j),
x(i,j,f)))))
                                  +
                                  (0.3*(sum((f),sum((i,j),
d(i,j)*x(i,j,f)))))
                                  +
                                  (0.4*(sum((f),sum((i,j),
bw(f)*x(i,j,f)))))  ;


source_node(f)             .. sum((j), x('n0',j,f)) =e= 1;

destination_node(f)     .. sum((i),x(i,'n13',f)) =e= 1;

intermediate1_node(f) .. sum((i,j), x(i,'n1',f)) -
sum((i,j), x('n1',j,f)) =e= 0;
intermediate2_node(f) .. sum((i,j), x(i,'n2',f)) -
sum((i,j), x('n2',j,f)) =e= 0;
intermediate3_node(f) .. sum((i,j), x(i,'n3',f)) -
sum((i,j), x('n3',j,f)) =e= 0;
```

```
intermediate4_node(f) .. sum((i,j), x(i,'n4',f)) - sum((i,j),
x('n4',j,f)) =e= 0;
intermediate5_node(f) .. sum((i,j), x(i,'n5',f)) - sum((i,j),
x('n5',j,f)) =e= 0;
intermediate6_node(f) .. sum((i,j), x(i,'n6',f)) - sum((i,j),
x('n6',j,f)) =e= 0;
intermediate7_node(f) .. sum((i,j), x(i,'n7',f)) - sum((i,j),
x('n7',j,f)) =e= 0;
intermediate8_node(f) .. sum((i,j), x(i,'n8',f)) - sum((i,j),
x('n8',j,f)) =e= 0;
intermediate9_node(f) .. sum((i,j), x(i,'n9',f)) - sum((i,j),
x('n9',j,f)) =e= 0;
intermediate10_node(f) .. sum((i,j), x(i,'n10',f)) -
sum((i,j), x('n10',j,f)) =e= 0;
intermediate10_node(f) .. sum((i,j), x(i,'n11',f)) -
sum((i,j), x('n11',j,f)) =e= 0;
intermediate12_node(f) .. sum((i,j), x(i,'n12',f)) -
sum((i,j), x('n12',j,f)) =e= 0;


Link_Capacity(i,j)     .. sum((f),bw(f)*x(i,j,f)) =l= c(i,j);


Model Transport /all/ ;


option mip=CPLEX
*option iterlim = 50000
*option profiletol = 1.0E-2
Solve transport using mip minimizing z ;


DL.l = sum((f),sum((i,j), d(i,j)*x.l(i,j,f)));
HC.l = sum((f),sum((i,j), x.l(i,j,f)));
CT.l = sum((f),sum((i,j), w(i,j)*x.l(i,j,f)));
BC.l = sum((f),sum((i,j), bw(f)*x.l(i,j,f)));


Display x.l
Display DL.l
Display HC.l
Display CT.l
Display BC.l
Display z.l
```

File in GAMS of Section 4.5.4 (Example 1)

```
************************************************
***                                        ***
***        Multi-Objective NSF Network      ***
***                                        ***
***        Objective Functions:             ***
***        Hop Count                        ***
***        Delay                            ***
***        Bandwidth Consumption            ***
***                                        ***
***     Eng. Yezid E. Donoso Meisel, Ph.D.  ***
************************************************
Sets
  i    ingress node / n0, n1, n2, n3, n4, n5, n6, n7, n8,
n9, n10, n11, n12, n13 /
  j    egress node  / n0, n1, n2, n3, n4, n5, n6, n7,
n8, n9, n10, n11, n12, n13 /
  f    Flows / f1 /
  t    destination nodes / n8, n12 / ;
Parameter
  bw(f) Transmission rate (Kbps)
        / f1 512 / ;
Table delay(i,j) Link Delay
```

	n0	n1	n2	n3	n4	n5	n6	n7	n8	n9	n10	n11	n12	n13
n0	999	9	9	7	999	999	999	999	999	999	999	999	999	999
n1	999	999	999	13	999	999	20	999	999	999	999	999	999	999
n2	999	999	999	999	7	999	999	2	999	999	999	999	999	999
n3	999	13	999	999	999	999	999	999	999	999	15	999	999	999
n4	999	999	7	999	999	7	999	999	999	999	11	999	999	999
n5	999	999	999	999	7	999	7	999	999	999	999	999	999	999
n6	999	20	999	999	999	7	999	999	999	7	999	999	999	999
n7	999	999	26	999	999	999	999	999	5	999	999	999	999	8
n8	999	999	999	999	999	999	999	5	999	5	999	999	7	999
n9	999	999	999	999	999	999	7	999	5	999	999	8	999	999
n10	999	999	999	15	11	999	999	999	999	999	999	9	14	999
n11	999	999	999	999	999	999	999	999	999	8	9	999	999	999
n12	999	999	999	999	999	999	999	999	7	999	14	999	999	4
n13	999	999	999	999	999	999	999	8	999	999	999	999	4	999 ;

```
Parameter d(i,j) delay ;
```

```
   d(i,j) = delay(i,j) ;
Table cap(i,j) Link Capacity (Kbps)
```

	n0	n1	n2	n3	n4	n5	n6	n7	n8	n9	n10	n11	n12	n13
n0	0	1536	1536	1536	0	0	0	0	0	0	0	0	0	0
n1	0	0	0	1536	0	0	1536	0	0	0	0	0	0	0
n2	0	0	0	0	1536	0	0	1536	0	0	0	0	0	0
n3	0	1536	0	0	0	0	0	0	0	0	1536	0	0	0
n4	0	0	1536	0	0	1536	0	0	0	0	1536	0	0	0
n5	0	0	0	0	1536	0	1536	0	0	0	0	0	0	0
n6	0	1536	0	0	0	1536	0	0	0	1536	0	0	0	0
n7	0	0	1536	0	0	0	0	0	1536	0	0	0	0	1536
n8	0	0	0	0	0	0	0	1536	0	1536	0	0	1536	0
n9	0	0	0	0	0	0	1536	0	1536	0	0	1536	0	0
n10	0	0	0	1536	1536	0	0	0	0	0	0	1536	1536	0
n11	0	0	0	0	0	0	0	0	0	1536	1536	0	0	0
n12	0	0	0	0	0	0	0	0	1536	0	1536	0	0	1536
n13	0	0	0	0	0	0	0	1536	0	0	0	0	1536	0 ;

```
Parameter c(i,j) Link Capacity ;
   c(i,j) = cap(i,j) ;
Table cost(i,j) Link Cost
```

	n0	n1	n2	n3	n4	n5	n6	n7	n8	n9	n10	n11	n12	n13
n0	999	9	9	7	999	999	999	999	999	999	999	999	999	999
n1	999	999	999	13	999	999	20	999	999	999	999	999	999	999
n2	999	999	999	999	7	999	999	26	999	999	999	999	999	999
n3	999	13	999	999	999	999	999	999	999	999	15	999	999	999
n4	999	999	7	999	999	7	999	999	999	999	11	999	999	999
n5	999	999	999	999	7	999	7	999	999	999	999	999	999	999
n6	999	20	999	999	999	7	999	999	999	7	999	999	999	999
n7	999	999	26	999	999	999	999	999	5	999	999	999	999	8
n8	999	999	999	999	999	999	999	5	999	5	999	999	7	999
n9	999	999	999	999	999	999	7	999	5	999	999	8	999	999
n10	999	999	999	15	11	999	999	999	999	999	999	9	14	999
n11	999	999	999	999	999	999	999	999	999	8	9	999	999	999
n12	999	999	999	999	999	999	999	999	7	999	14	999	999	4
n13	999	999	999	999	999	999	999	8	999	999	999	999	4	999 ;

```
Parameter w(i,j) cost ;
   w(i,j) = cost(i,j) ;
Variables
   x(i,j,t,f)  If the link i j is used (0) to the flow f
with destination node t
   HC    Hop Count (Hops)
   DL    Delay (ms)
```

```
   CT    Cost ($)
   BC    Bandwidth Consumption (Kbps)
   r1    r1
   r2    r2
   z     minimizacion ;
Positive Variable x;
Equations
Shortest_Path            Objective function
source_node(t,f)                   source node
destination_node_8(f)              destination node
destination_node_12(f)             destination node
intermediate1_node(t,f)     intermediate node
intermediate2_node(t,f)     intermediate node
intermediate3_node(t,f)     intermediate node
intermediate4_node(t,f)     intermediate node
intermediate5_node(t,f)     intermediate node
intermediate6_node(t,f)     intermediate node
intermediate7_node(t,f)     intermediate node
intermediate9_node(t,f)     intermediate node
intermediate10_node(t,f)     intermediate node
intermediate11_node(t,f)     intermediate node
intermediate13_node(t,f)     intermediate node
Link_capacity(i,j)    Link Capacity ;
;


Shortest_Path.. z =e= (0.5*(sum((f), (sum((t), sum((i,j),
x(i,j,t,f)))))))
                        +
                        (0.5*(sum((f), (sum((t), sum((i,j),
d(i,j)*x(i,j,t,f)))))))
                        +
                        (0.0*(sum((f), sum((i,j),
bw(f)*smax((t),x(i,j,t,f))))))))   ;

source_node(t,f)       .. sum((j), x('n0',j,t,f)) =e= 1;

destination_node_8(f) .. sum((i),x(i,'n8','n8',f)) =e= 1;
destination_node_12(f) .. sum((i),x(i,'n12','n12',f)) =e= 1;

intermediate1_node(t,f) .. sum((i,j), x(i,'n1',t,f)) -
sum((i,j), x('n1',j,t,f)) =e= 0;
```

```
intermediate2_node(t,f) .. sum((i,j), x(i,'n2',t,f)) -
sum((i,j), x('n2',j,t,f)) =e= 0;
intermediate3_node(t,f) .. sum((i,j), x(i,'n3',t,f)) -
sum((i,j), x('n3',j,t,f)) =e= 0;
intermediate4_node(t,f) .. sum((i,j), x(i,'n4',t,f)) -
sum((i,j), x('n4',j,t,f)) =e= 0;
intermediate5_node(t,f) .. sum((i,j), x(i,'n5',t,f)) -
sum((i,j), x('n5',j,t,f)) =e= 0;
intermediate6_node(t,f) .. sum((i,j), x(i,'n6',t,f)) -
sum((i,j), x('n6',j,t,f)) =e= 0;
intermediate7_node(t,f) .. sum((i,j), x(i,'n7',t,f)) -
sum((i,j), x('n7',j,t,f)) =e= 0;
intermediate9_node(t,f) .. sum((i,j), x(i,'n9',t,f)) -
sum((i,j), x('n9',j,t,f)) =e= 0;
intermediate10_node(t,f) .. sum((i,j), x(i,'n10',t,f)) -
sum((i,j), x('n10',j,t,f)) =e= 0;
intermediate11_node(t,f) .. sum((i,j), x(i,'n11',t,f)) -
sum((i,j), x('n11',j,t,f)) =e= 0;
intermediate13_node(t,f) .. sum((i,j), x(i,'n13',t,f)) -
sum((i,j), x('n13',j,t,f)) =e= 0;


Link_Capacity(i,j) .. sum((f),bw(f)*smax((t),x(i,j,t,f))) =l=
c(i,j);

Model Transport /all/ ;

*option dnlp=CONOPT
option dnlp=SNOPT
*option iterlim = 50000
*option profiletol = 1.0E-2
solve transport using dnlp minimizing z ;

DL.l = sum((f), sum((t), sum((i,j), d(i,j)*x.l(i,j,t,f))));
HC.l = sum((f), sum((t), sum((i,j), x.l(i,j,t,f))));
CT.l = sum((f), sum((t), sum((i,j), w(i,j)*x.l(i,j,t,f))));
BC.l = sum((f), sum((i,j), bw(f)*smax((t),x.l(i,j,t,f))));

Display x.l
Display DL.l
Display HC.l
Display CT.l
```

```
Display BC.l
Display z.l
```

File in GAMS of Section 4.5.4 (Example 2)

```
******************************************************
***                                                ***
***        Multi-Objective NSF Network             ***
***                                                ***
***        Objective Functions:                    ***
***        Hop Count                               ***
***        Delay                                   ***
***        Bandwidth Consumption                   ***
***                                                ***
***      Eng. Yezid E. Donoso Meisel, Ph.D.        ***
******************************************************
Sets
  i   ingress node / n0, n1, n2, n3, n4, n5, n6, n7, n8,
n9, n10, n11, n12, n13 /
  j   egress node  / n0, n1, n2, n3, n4, n5, n6, n7,
n8, n9, n10, n11, n12, n13 /
  f   Flows / f1, f2 /
  t   destination nodes / n8, n12 / ;
Parameter
  bw(f) Transmission rate (Kbps)
        / f1 1024
          f2 1024 / ;
```

```
Table delay(i,j) Link Delay
```

	n0	n1	n2	n3	n4	n5	n6	n7	n8	n9	n10	n11	n12	n13
n0	999	9	9	7	999	999	999	999	999	999	999	999	999	999
n1	999	999	999	13	999	999	20	999	999	999	999	999	999	999
n2	999	999	999	999	7	999	999	26	999	999	999	999	999	999
n3	999	13	999	999	999	999	999	999	999	999	15	999	999	999
n4	999	999	7	999	999	7	999	999	999	999	11	999	999	999
n5	999	999	999	999	7	999	7	999	999	999	999	999	999	999
n6	999	20	999	999	999	7	999	999	999	7	999	999	999	999
n7	999	999	26	999	999	999	999	999	5	999	999	999	999	8
n8	999	999	999	999	999	999	999	5	999	5	999	999	7	999
n9	999	999	999	999	999	999	7	999	5	999	999	8	999	999
n10	999	999	999	15	11	999	999	999	999	999	999	9	14	999
n11	999	999	999	999	999	999	999	999	999	8	9	999	999	999
n12	999	999	999	999	999	999	999	999	7	999	14	999	999	4
n13	999	999	999	999	999	999	999	8	999	999	999	999	4	999 ;

```
Parameter d(i,j) delay ;
  d(i,j) = delay(i,j) ;
Table cap(i,j) Link Capacity (Kbps)
```

	n0	n1	n2	n3	n4	n5	n6	n7	n8	n9	n10	n11	n12	n13
n0	0	1536	1536	1536	0	0	0	0	0	0	0	0	0	0
n1	0	0	0	1536	0	0	1536	0	0	0	0	0	0	0
n2	0	0	0	0	1536	0	0	1536	0	0	0	0	0	0
n3	0	1536	0	0	0	0	0	0	0	0	1536	0	0	0
n4	0	0	1536	0	0	1536	0	0	0	0	1536	0	0	0
n5	0	0	0	0	1536	0	1536	0	0	0	0	0	0	0
n6	0	1536	0	0	0	1536	0	0	0	1536	0	0	0	0
n7	0	0	1536	0	0	0	0	0	1536	0	0	0	0	1536
n8	0	0	0	0	0	0	0	1536	0	1536	0	0	1536	0
n9	0	0	0	0	0	0	1536	0	1536	0	0	1536	0	0
n10	0	0	0	1536	1536	0	0	0	0	0	0	1536	1536	0
n11	0	0	0	0	0	0	0	0	0	1536	1536	0	0	0
n12	0	0	0	0	0	0	0	0	1536	0	1536	0	0	1536
n13	0	0	0	0	0	0	0	1536	0	0	0	0	1536	0 ;

```
Parameter c(i,j) Link Capacity ;
  c(i,j) = cap(i,j) ;
```

Table cost(i,j) Link Cost

	n0	n1	n2	n3	n4	n5	n6	n7	n8	n9	n10	n11	n12	n13
n0	999	9	9	7	999	999	999	999	999	999	999	999	999	999
n1	999	999	999	13	999	999	20	999	999	999	999	999	999	999
n2	999	999	999	999	7	999	999	26	999	999	999	999	999	999
n3	999	13	999	999	999	999	999	999	999	999	15	999	999	999
n4	999	999	7	999	999	7	999	999	999	999	11	999	999	999
n5	999	999	999	999	7	999	7	999	999	999	999	999	999	999
n6	999	20	999	999	999	7	999	999	999	7	999	999	999	999
n7	999	999	26	999	999	999	999	999	5	999	999	999	999	8
n8	999	999	999	999	999	999	999	5	999	5	999	999	7	999
n9	999	999	999	999	999	999	7	999	5	999	999	8	999	999
n10	999	999	999	15	11	999	999	999	999	999	999	9	14	999
n11	999	999	999	999	999	999	999	999	999	8	9	999	999	999
n12	999	999	999	999	999	999	999	999	7	999	14	999	999	4
n13	999	999	999	999	999	999	999	8	999	999	999	999	4	999;

```
Parameter w(i,j) cost ;
  w(i,j) = cost(i,j) ;
Variables
  x(i,j,t,f)  If the link i j is used (0) to the flow f
with destination node t
  HC      Hop Count (Hops)
  DL      Delay (ms)
  CT      Cost ($)
  BC      Bandwidth Consumption (Kbps)
  r1      r1
  r2      r2
  z       minimization ;
Positive Variable x;
Equations
Shortest_Path          Objective function
source_node(t,f)             source node
destination_node_8(f)          destination node
destination_node_12(f)          destination node
intermediate1_node(t,f)     intermediate node
intermediate2_node(t,f)     intermediate node
intermediate3_node(t,f)     intermediate node
intermediate4_node(t,f)     intermediate node
intermediate5_node(t,f)     intermediate node
intermediate6_node(t,f)     intermediate node
intermediate7_node(t,f)     intermediate node
```

```
intermediate9_node(t,f)        intermediate node
intermediate10_node(t,f)       intermediate node
intermediate11_node(t,f)       intermediate node
intermediate13_node(t,f)       intermediate node
Link_capacity(i,j)    Link Capacity ;
;


Shortest_Path  ..  z =e= (0.1*(sum((f), (sum((t), sum((i,j),
x(i,j,t,f)))))))
                          +
                     (0.2*(sum((f), (sum((t), sum((i,j),
d(i,j)*x(i,j,t,f)))))))
                          +
                     (0.7*(sum((f), sum((i,j),
bw(f)*smax((t),x(i,j,t,f))))))  ;


source_node(t,f)  ..  sum((j), x('n0',j,t,f)) =e= 1;


destination_node_8(f)  ..  sum((i),x(i,'n8','n8',f)) =e= 1;
destination_node_12(f)  ..  sum((i),x(i,'n12','n12',f)) =e= 1;


intermediate1_node(t,f)  ..  sum((i,j), x(i,'n1',t,f)) -
sum((i,j), x('n1',j,t,f)) =e= 0;
intermediate2_node(t,f)  ..  sum((i,j), x(i,'n2',t,f)) -
sum((i,j), x('n2',j,t,f)) =e= 0;
intermediate3_node(t,f)  ..  sum((i,j), x(i,'n3',t,f)) -
sum((i,j), x('n3',j,t,f)) =e= 0;
intermediate4_node(t,f)  ..  sum((i,j), x(i,'n4',t,f)) -
sum((i,j), x('n4',j,t,f)) =e= 0;
intermediate5_node(t,f)  ..  sum((i,j), x(i,'n5',t,f)) -
sum((i,j), x('n5',j,t,f)) =e= 0;
intermediate6_node(t,f)  ..  sum((i,j), x(i,'n6',t,f)) -
sum((i,j), x('n6',j,t,f)) =e= 0;
intermediate7_node(t,f)  ..  sum((i,j), x(i,'n7',t,f)) -
sum((i,j), x('n7',j,t,f)) =e= 0;
intermediate9_node(t,f)  ..  sum((i,j), x(i,'n9',t,f)) -
sum((i,j), x('n9',j,t,f)) =e= 0;
intermediate10_node(t,f)..sum((i,j), x(i,'n10',t,f)) -
sum((i,j), x('n10',j,t,f)) =e= 0;
intermediate11_node(t,f)..sum((i,j), x(i,'n11',t,f)) -
sum((i,j), x('n11',j,t,f)) =e= 0;
intermediate13_node(t,f)..sum((i,j), x(i,'n13',t,f)) -
sum((i,j), x('n13',j,t,f)) =e= 0;
```

```
Link_Capacity(i,j)..sum((f),bw(f)*smax((t),x(i,j,t,f)))
=l= c(i,j);

Model Transport /all/ ;

option dnlp=CONOPT
*option dnlp=SNOPT
*option iterlim = 50000
*option profiletol = 1.0E-2
solve transport using dnlp minimizing z ;

DL.l = sum((f), sum((t), sum((i,j), d(i,j)*x.l(i,j,t,f))));
HC.l = sum((f), sum((t), sum((i,j), x.l(i,j,t,f))));
CT.l = sum((f), sum((t), sum((i,j), w(i,j)*x.l(i,j,t,f))));
BC.l = sum((f), sum((i,j), bw(f)*smax((t),x.l(i,j,t,f))));

Display x.l
Display DL.l
Display HC.l
Display CT.l
Display BC.l
Display z.l
```

File in GAMS of Section 4.5.5

```
****************************************************
***                                            ***
***        Multi-Objective NSF Network          ***
***                                            ***
***        Objective Functions:                 ***
***        Hop Count                            ***
***        Delay                                ***
***        Bandwidth Consumption                ***
***        Maximum Link Utilization             ***
***                                            ***
***        Eng. Yezid E. Donoso Meisel, Ph.D.   ***
****************************************************
```

```
Sets
   i    ingress node / n0, n1, n2, n3, n4, n5, n6, n7, n8,
n9, n10, n11, n12, n13 /
   j    egress node  / n0, n1, n2, n3, n4, n5, n6, n7,
n8, n9, n10, n11, n12, n13 /
   f    Flows / f1, f2, f3 / ;
Parameter
   bw(f) Transmission rate (Kbps)
        / f1 1024
          f2 1024
          f3 1024/ ;
Table delay(i,j) Link Delay
```

	n0	n1	n2	n3	n4	n5	n6	n7	n8	n9	n10	n11	n12	n13
n0	999	9	9	7	999	999	999	999	999	999	999	999	999	999
n1	999	999	999	13	999	999	20	999	999	999	999	999	999	999
n2	999	999	999	999	7	999	999	26	999	999	999	999	999	999
n3	999	13	999	999	999	999	999	999	999	999	15	999	999	999
n4	999	999	7	999	999	7	999	999	999	999	11	999	999	999
n5	999	999	999	999	7	999	7	999	999	999	999	999	999	999
n6	999	20	999	999	999	7	999	999	999	7	999	999	999	999
n7	999	999	26	999	999	999	999	999	5	999	999	999	999	8
n8	999	999	999	999	999	999	999	5	999	5	999	999	7	999
n9	999	999	999	999	999	999	7	999	5	999	999	8	999	999
n10	999	999	999	15	11	999	999	999	999	999	999	9	14	999
n11	999	999	999	999	999	999	999	999	999	8	9	999	999	999
n12	999	999	999	999	999	999	999	999	7	999	14	999	999	4
n13	999	999	999	999	999	999	999	8	999	999	999	999	4	999 ;

```
Parameter d(i,j) delay ;
   d(i,j) = delay(i,j) ;
```

```
Table cap(i,j)  Link Capacity (Kbps)

     n0  n1    n2    n3    n4    n5    n6    n7    n8    n9    n10   n11   n12   n13
n0   0   1536  1536  1536  0     0     0     0     0     0     0     0     0     0
n1   0   0     0     1536  0     0     1536  0     0     0     0     0     0     0
n2   0   0     0     0     1536  0     0     1536  0     0     0     0     0     0
n3   0   1536  0     0     0     0     0     0     0     0     1536  0     0     0
n4   0   0     1536  0     0     1536  0     0     0     0     1536  0     0     0
n5   0   0     0     0     1536  0     1536  0     0     0     0     0     0     0
n6   0   1536  0     0     0     1536  0     0     0     1536  0     0     0     0
n7   0   0     1536  0     0     0     0     0     1536  0     0     0     0     1536
n8   0   0     0     0     0     0     0     1536  0     1536  0     0     1536  0
n9   0   0     0     0     0     0     1536  0     1536  0     0     1536  0     0
n10  0   0     0     1536  1536  0     0     0     0     0     0     1536  1536  0
n11  0   0     0     0     0     0     0     0     0     1536  1536  0     0     0
n12  0   0     0     0     0     0     0     0     1536  0     1536  0     0     1536
n13  0   0     0     0     0     0     0     1536  0     0     0     0     1536  0 ;

Parameter c(i,j)  Link Capacity ;
   c(i,j) = cap(i,j) ;
Table cost(i,j)  Link Cost

     n0   n1   n2   n3   n4   n5   n6   n7   n8   n9   n10  n11  n12  n13
n0   999  9    9    7    999  999  999  999  999  999  999  999  999  999
n1   999  999  999  13   999  999  20   999  999  999  999  999  999  999
n2   999  999  999  999  7    999  999  26   999  999  999  999  999  999
n3   999  13   999  999  999  999  999  999  999  999  15   999  999  999
n4   999  999  7    999  999  7    999  999  999  999  11   999  999  999
n5   999  999  999  999  7    999  7    999  999  999  999  999  999  999
n6   999  20   999  999  999  7    999  999  999  7    999  999  999  999
n7   999  999  26   999  999  999  999  999  5    999  999  999  999  8
n8   999  999  999  999  999  999  999  5    999  5    999  999  7    999
n9   999  999  999  999  999  999  7    999  5    999  999  8    999  999
n10  999  999  999  15   11   999  999  999  999  999  999  9    14   999
n11  999  999  999  999  999  999  999  999  999  8    9    999  999  999
n12  999  999  999  999  999  999  999  999  7    999  14   999  999  4
n13  999  999  999  999  999  999  999  8    999  999  999  999  4    999 ;

Parameter w(i,j) cost ;
   w(i,j) = cost(i,j) ;
Variables
   x(i,j,f)    If the link i j is used (0) to the flow f
   HC      Hop Count (Hops)
   DL      Delay (ms)
   CT      Cost ($)
   BC      Bandwidth Consumption (Kbps)
```

```
   a        MLU Value
   alpha(i,j)  MLU Value in each link
   z         minimization ;
Positive Variable x;
Equations
Shortest_Path          Objective function
MLU_Link(i,j)            MLU Link Value
MLU                      MLU Total Value
source_node(f)           source node
destination_node(f)         destination node
intermediate1_node(f)    intermediate node
intermediate2_node(f)    intermediate node
intermediate3_node(f)    intermediate node
intermediate4_node(f)    intermediate node
intermediate5_node(f)    intermediate node
intermediate6_node(f)    intermediate node
intermediate7_node(f)    intermediate node
intermediate8_node(f)    intermediate node
intermediate9_node(f)    intermediate node
intermediate10_node(f)    intermediate node
intermediate11_node(f)    intermediate node
intermediate12_node(f)    intermediate node
Link_capacity(i,j)   Link Capacity ;

Shortest_Path .. z =e= (1.0*(sum((f),sum((i,j),
ceil(x(i,j,f))))))
                            +
                            (0.0*(sum((f),sum((i,j),
d(i,j)*ceil(x(i,j,f)))))))
                            +
                            (0.0*(sum((f),sum((i,j),
bw(f)*x(i,j,f)))))
                            +
                            (0.0 * a) ;

MLU_Link(i,j)          .. c(i,j)*alpha(i,j) =e= (sum((f),
bw(f)*x(i,j,f)));

MLU                    .. a =e= smax((i,j), alpha(i,j));
```

```
source_node(f)  ..  sum((j),  x('n0',j,f))  =e=  1;

destination_node(f)  ..  sum((i),x(i,'n13',f))  =e=  1;

intermediate1_node(f)  ..  sum((i,j),  x(i,'n1',f))  -
sum((i,j),  x('n1',j,f))  =e=  0;
intermediate2_node(f)  ..  sum((i,j),  x(i,'n2',f))  -
sum((i,j),  x('n2',j,f))  =e=  0;
intermediate3_node(f)  ..  sum((i,j),  x(i,'n3',f))  -
sum((i,j),  x('n3',j,f))  =e=  0;
intermediate4_node(f)  ..  sum((i,j),  x(i,'n4',f))  -
sum((i,j),  x('n4',j,f))  =e=  0;
intermediate5_node(f)  ..  sum((i,j),  x(i,'n5',f))  -
sum((i,j),  x('n5',j,f))  =e=  0;
intermediate6_node(f)  ..  sum((i,j),  x(i,'n6',f))  -
sum((i,j),  x('n6',j,f))  =e=  0;
intermediate7_node(f)  ..  sum((i,j),  x(i,'n7',f))  -
sum((i,j),  x('n7',j,f))  =e=  0;
intermediate8_node(f)  ..  sum((i,j),  x(i,'n8',f))  -
sum((i,j),  x('n8',j,f))  =e=  0;
intermediate9_node(f)  ..  sum((i,j),  x(i,'n9',f))  -
sum((i,j),  x('n9',j,f))  =e=  0;
intermediate10_node(f)  ..  sum((i,j),  x(i,'n10',f))  -
sum((i,j),  x('n10',j,f))  =e=  0;
intermediate11_node(f)  ..  sum((i,j),  x(i,'n11',f))  -
sum((i,j),  x('n11',j,f))  =e=  0;
intermediate12_node(f)  ..  sum((i,j),  x(i,'n12',f))  -
sum((i,j),  x('n12',j,f))  =e=  0;

Link_Capacity(i,j)  ..sum((f),bw(f)*x(i,j,f))  =l=  c(i,j);



Model Transport /all/ ;

option dnlp=SNOPT
*option iterlim = 50000
*option profiletol = 1.0E-2
Solve transport using dnlp minimizing z ;

DL.l = sum((f),sum((i,j),  d(i,j)*ceil(x.l(i,j,f))));
HC.l = sum((f),sum((i,j),  ceil(x.l(i,j,f))));
CT.l = sum((f),sum((i,j),  w(i,j)*x.l(i,j,f)));
```

```
BC.l = sum((f),sum((i,j), bw(f)*x.l(i,j,f)));

Display x.l
Display DL.l
Display HC.l
Display CT.l
Display BC.l
Display a.l
Display z.l
```

File in GAMS of Section 4.5.6

```
***********************************************************
***                                                     ***
***          Multi-Objective        Network             ***
***                                                     ***
***          Objective Functions:                       ***
***          Hop Count                                  ***
***          Delay                                      ***
***          Bandwidth Consumption                      ***
***          Maximum Link Utilization                   ***
***                                                     ***
***          Eng. Yezid E. Donoso Meisel, Ph.D.         ***
***********************************************************
Sets
  i    ingress node / n0, n1, n2, n3, n4, n5, n6, n7, n8,
n9, n10, n11, n12, n13 /
  j    egress node  / n0, n1, n2, n3, n4, n5, n6, n7,
n8, n9, n10, n11, n12, n13 /
  f    Flows / f1 /
  t    destination nodes / n8, n12 / ;
Parameter
  bw(f) Transmission rate (Kbps)
        / f1 2048  /     ;
```

Table delay(i,j) Link Delay

	n0	n1	n2	n3	n4	n5	n6	n7	n8	n9	n10	n11	n12	n13
n0	999	9	9	7	999	999	999	999	999	999	999	999	999	999
n1	999	999	999	13	999	999	20	999	999	999	999	999	999	999
n2	999	999	999	999	7	999	999	26	999	999	999	999	999	999
n3	999	13	999	999	999	999	999	999	999	999	15	999	999	999
n4	999	999	7	999	999	7	999	999	999	999	11	999	999	999
n5	999	999	999	999	7	999	7	999	999	999	999	999	999	999
n6	999	20	999	999	999	7	999	999	999	7	999	999	999	999
n7	999	999	26	999	999	999	999	999	5	999	999	999	999	8
n8	999	999	999	999	999	999	999	5	999	5	999	999	7	999
n9	999	999	999	999	999	999	7	999	5	999	999	8	999	999
n10	999	999	999	15	11	999	999	999	999	999	999	9	14	999
n11	999	999	999	999	999	999	999	999	999	8	9	999	999	999
n12	999	999	999	999	999	999	999	999	7	999	14	999	999	4
n13	999	999	999	999	999	999	999	8	999	999	999	999	4	999 ;

Parameter d(i,j) delay ;
  d(i,j) = delay(i,j) ;
Table cap(i,j) Link Capacity (Kbps)

	n0	n1	n2	n3	n4	n5	n6	n7	n8	n9	n10	n11	n12	n13
n0	0	1536	1536	1536	0	0	0	0	0	0	0	0	0	0
n1	0	0	0	1536	0	0	1536	0	0	0	0	0	0	0
n2	0	0	0	0	1536	0	0	1536	0	0	0	0	0	0
n3	0	1536	0	0	0	0	0	0	0	0	1536	0	0	0
n4	0	0	1536	0	0	1536	0	0	0	0	1536	0	0	0
n5	0	0	0	0	1536	0	1536	0	0	0	0	0	0	0
n6	0	1536	0	0	0	1536	0	0	0	1536	0	0	0	0
n7	0	0	1536	0	0	0	0	0	1536	0	0	0	0	1536
n8	0	0	0	0	0	0	0	1536	0	1536	0	0	1536	0
n9	0	0	0	0	0	0	1536	0	1536	0	0	1536	0	0
n10	0	0	0	1536	1536	0	0	0	0	0	0	1536	1536	0
n11	0	0	0	0	0	0	0	0	0	1536	1536	0	0	0
n12	0	0	0	0	0	0	0	0	1536	0	1536	0	0	1536
n13	0	0	0	0	0	0	0	1536	0	0	0	0	1536	0 ;

Parameter c(i,j) Link Capacity ;
  c(i,j) = cap(i,j) ;

```
Table cost(i,j) Link Cost

     n0   n1   n2   n3   n4   n5   n6   n7   n8   n9  n10  n11  n12  n13
n0  999    9    9    7  999  999  999  999  999  999  999  999  999  999
n1  999  999  999   13  999  999   20  999  999  999  999  999  999  999
n2  999  999  999  999    7  999  999   26  999  999  999  999  999  999
n3  999   13  999  999  999  999  999  999  999  999   15  999  999  999
n4  999  999    7  999  999    7  999  999  999  999   11  999  999  999
n5  999  999  999  999    7  999    7  999  999  999  999  999  999  999
n6  999   20  999  999  999    7  999  999  999    7  999  999  999  999
n7  999  999   26  999  999  999  999  999    5  999  999  999  999    8
n8  999  999  999  999  999  999  999    5  999    5  999  999    7  999
n9  999  999  999  999  999  999    7  999    5  999  999    8  999  999
n10 999  999  999   15   11  999  999  999  999  999  999    9   14  999
n11 999  999  999  999  999  999  999  999  999    8    9  999  999  999
n12 999  999  999  999  999  999  999  999    7  999   14  999  999    4
n13 999  999  999  999  999  999  999    8  999  999  999  999    4  999 ;

Parameter w(i,j)  cost ;
   w(i,j) = cost(i,j) ;
Variables
   x(i,j,t,f)  If the link i j is used (0) to the flow f
with destination node t
   HC      Hop Count (Hops)
   DL      Delay (ms)
   CT      Cost ($)
   BC      Bandwidth Consumption (Kbps)
   a       MLU Value
   alpha(i,j)  MLU Value in each link
   z       minimization ;
Positive Variable x;
Equations
Shortest_Path           Objective function
MLU_Link(i,j)             MLU Link Value
MLU                       MLU Total Value
source_node(t,f)          source node
destination_node_8(f)       destination node
destination_node_12(f)       destination node
intermediate1_node(t,f)    intermediate node
intermediate2_node(t,f)    intermediate node
intermediate3_node(t,f)    intermediate node
intermediate4_node(t,f)    intermediate node
intermediate5_node(t,f)    intermediate node
```

```
intermediate6_node(t,f)         intermediate node
intermediate7_node(t,f)         intermediate node
intermediate9_node(t,f)         intermediate node
intermediate10_node(t,f)        intermediate node
intermediate11_node(t,f)        intermediate node
intermediate13_node(t,f)        intermediate node
Link_capacity(i,j)    Link Capacity ;
;


Shortest_Path            .. z =e= (0.1*(sum((f), (sum((t),
sum((i,j), ceil(x(i,j,t,f)))))))))
                                +
                        (0.1*(sum((f), (sum((t),
sum((i,j), d(i,j)*ceil(x(i,j,t,f))))))))
                                +
                        (0.1*(sum((f), sum((i,j),
bw(f)*smax((t),x(i,j,t,f))))))
                                +
                        (0.7*a)  ;


MLU_Link(i,j)      .. c(i,j)*alpha(i,j) =e= (sum((f),
bw(f)*smax((t),x(i,j,t,f))));


MLU                .. a =e= smax((i,j), alpha(i,j));



source_node(t,f)  .. sum((j), x('n0',j,t,f)) =e= 1;

destination_node_8(f) .. sum((i),x(i,'n8','n8',f)) =e= 1;
destination_node_12(f) .. sum((i),x(i,'n12','n12',f)) =e=
1;

intermediate1_node(t,f) .. sum((i,j), x(i,'n1',t,f)) -
sum((i,j), x('n1',j,t,f)) =e= 0;
intermediate2_node(t,f) .. sum((i,j), x(i,'n2',t,f)) -
sum((i,j), x('n2',j,t,f)) =e= 0;
intermediate3_node(t,f) .. sum((i,j), x(i,'n3',t,f)) -
sum((i,j), x('n3',j,t,f)) =e= 0;
intermediate4_node(t,f) .. sum((i,j), x(i,'n4',t,f)) -
sum((i,j), x('n4',j,t,f)) =e= 0;
intermediate5_node(t,f) .. sum((i,j), x(i,'n5',t,f)) -
sum((i,j), x('n5',j,t,f)) =e= 0;
```

```
intermediate6_node(t,f)  .. sum((i,j), x(i,'n6',t,f)) -
sum((i,j), x('n6',j,t,f)) =e= 0;
intermediate7_node(t,f)  .. sum((i,j), x(i,'n7',t,f)) -
sum((i,j), x('n7',j,t,f)) =e= 0;
intermediate9_node(t,f)  .. sum((i,j), x(i,'n9',t,f)) -
sum((i,j), x('n9',j,t,f)) =e= 0;
intermediate10_node (t,f)  .. sum((i,j), x(i,'n10',t,f)) -
sum((i,j), x('n10',j,t,f)) =e= 0;
intermediate11_node(t,f)  .. sum((i,j), x(i,'n11',t,f)) -
sum((i,j), x('n11',j,t,f)) =e= 0;
intermediate13_node(t,f)  .. sum((i,j), x(i,'n13',t,f)) -
sum((i,j), x('n13',j,t,f)) =e= 0;


Link_Capacity(i,j)       ..
sum((f),bw(f)*smax((t),x(i,j,t,f))) =l= c(i,j);

Model Transport /all/ ;

option dnlp=CONOPT
*option dnlp=SNOPT
*option iterlim = 50000
*option profiletol = 1.0E-2
solve transport using dnlp minimizing z ;

DL.l = sum((f), sum((t), sum((i,j),
d(i,j)*ceil(x.l(i,j,t,f)))));
HC.l = sum((f), sum((t), sum((i,j), ceil(x.l(i,j,t,f)))));
CT.l = sum((f), sum((t), sum((i,j), w(i,j)*x.l(i,j,t,f))));
BC.l = sum((f), sum((i,j), bw(f)*smax((t),x.l(i,j,t,f))));

Display x.l
Display DL.l
Display HC.l
Display CT.l
Display BC.l
Display a.l
Display z.l
```

Annex B[1]

Program in C to Section 4.7.1

```
/-------------------------------------------------------
/*


Program: MOEA - UNICAST


*/


#include <vcl.h>
#pragma hdrstop


#include <stdlib.h>
#include <math.h>
#include <stdio.h>
#include <conio.h>
#include <time.h>


//-------------------------------------------------------
//Basic Definitions
#define WHITE 0
#define GRAY 1
#define BLACK 2
#define MAX_NODES 50
```

---

[1] These source files are available online at http://www.crcpress.com/e_products/
downloads/default.asp

```
#define MAX_POPULATION 50
#define MAX_GENERATIONS 1
#define SIZE_COMPARISON 1.0
#define CROSS_OVER_PROBABILITY 0.75
#define MUTATION_PROBABILITY 0.1
#define ERROR 0.00001
#define oo 1000000000
#define oof 10000.0
#define zero 0.000001
#define DOMINATED 0
#define NONDOMINATED 1
#define BOTH_NONDOMINATED 2
#define MAX_NONDOMINATED_SOLUTIONS 3
//End basic declaration

//Declarations
float Prob[100][4];        // for roulette selection
float Jprob[100][3];       // for roulette selection
int num_prob;              // for roulette selection
float Ci[100][9];          // for roulette selection
int clusters[100][500];    // Needed for Clustering
int n;   // number of nodes
int e;   // number of edges
int d;   // number of sinks
int demand; //demand of the flow
int num_paths; //number of paths in the paths array
int max_paths; //indicate the destination node with maximum
number of paths
int num_population; //number of elements of the population
int num_population_initial  ; //number of elements of the
population
int new_num_population; //number of elements of the
population
int new_num_population1; //number of elements of the
population
int new_num_population2; //number of elements of the
population
int new_num_population3; //number of elements of the
population
int size_comparison_set; //size of the comparison set
int maxNondominated;// maximum size of external
nondominated set
```

```
int f1_max_value,f1_min_value;//Are the bounds of the
function f1
int f2_max_value,f2_min_value;//Are the bounds of the
function f2


int del_path[MAX_NODES];//path to be removed by a constraint
int color[MAX_NODES]; // needed for breadth-first search
int pred[MAX_NODES];   // array to store augmenting path
int dest[MAX_NODES];   // array to store the sink nodes
float strength[MAX_NODES][2];   // array to store the
strength
int capacity[MAX_NODES][MAX_NODES];  //array to store the
link capacity
int delay[MAX_NODES][MAX_NODES];   //array to store the
link delay
int visited[MAX_NODES][MAX_NODES];   //array to store
information about the link visited in the paths search
int num_paths_by_destination[MAX_NODES][3]; //array to
indicate the number of paths by each destination
                                 // 0 : destination
node
                                             // 1 :
number of paths to this destination
                                             // 2 :
number of the row where is the first position for
this destination
int paths[MAX_NODES][MAX_NODES]; //array to store the paths
to all destinations
                          // 0 : destination node
                          // 1 to (-1) : nodes of this path
                          // with (-1) values identify the
end of this path
int population[MAX_POPULATION][MAX_POPULATION];   //array
to store the population
int new_population[MAX_POPULATION][MAX_POPULATION];
//array to store the population
int new_population1[MAX_POPULATION][MAX_POPULATION];
//array to store the population for binary tournament
selection
int new_population2[MAX_POPULATION][MAX_POPULATION];
//array to store the population
```

```
int new_population3[MAX_POPULATION][MAX_POPULATION];
//array to store the population
int temp_population[MAX_POPULATION][MAX_POPULATION];
float fitness[MAX_POPULATION][2];//array to store fitness
of the chromosomes of the population
                                      // 0 : hop count
                                      // 1 : delay

int comparison_set[MAX_POPULATION][MAX_POPULATION];
//array to store the comparison set
int count_generations; //Generation Counter

clock_t time1, time2, time3;
//A Queue for Breadth-First Search--------------------
int head,tail;
int q[MAX_NODES+2];

void enqueue (int x) {
    q[tail] = x;
    tail++;
    color[x] = GRAY;
}
int dequeue () {
    int x = q[head];
    head++;
    color[x] = BLACK;
    return x;
}
//------------------------------------------------------------
//Breadth-First Search----------------------------------
int bfs (int start, int target) {
    int u,v;
    for (u=0; u<n; u++) {
        color[u] = WHITE;
    }
    head = tail = 0;
    enqueue(start);
    pred[start] = -1;
    while (head!=tail) {
        u = dequeue();
    // Search all adjacent white nodes v. If the capacity
```

```
        // from u to v in the residual network is positive,
        // enqueue v.
        for (v=0; v<n; v++) {
            if (color[v]==WHITE && visited[u][v] == 1) {
                enqueue(v);
                pred[v] = u;
            }
        }
    }

    // If the color of the target node is black now,
    // it means that we reached it.
    return color[target]==BLACK;
}
  int ai=0,aj=0;
//End Breadth First Search-------------------------------
#define MAX_PATHS 500 //Needed for backtracking generation
paths
int s=0,ind=0,kk=1,target,a=1,kn=0; //Needed for back-
tracking generation paths
int path[MAX_PATHS][MAX_NODES],ant[50][2]; //Needed for
backtracking generation paths

//-------Search Paths for Initial Population-----------
void update_population(){ //Update Backtracking population
        int w=0;
        while(path[num_population_initial   -1][w]!=-1){
            population[num_population_initial-
              1][w]=path[num_population_initial-1][w];
            w++;
        }
       if(path[num_population_initial-1][w]==-1){
            population[num_population_initial-
1][w]=path[num_population_initial-1][w];
        }
  }
//-----------------------------------------------------------
void move(int ii)
{   int u;
  for(u=ii;u<kk;u++)
  {
        path[num_paths-1][u]=path[num_paths-1][u+1];
```

```
    }
        path[num_paths-1][u]=-1;
}
//------------------------------------------------------
void del_visited(int base){ //Analyze visited nodes

  for(int ii=kk;ii>0;ii--)
  {
    if(visited[path[num_paths-1][ii-1]][path[num_paths-
      1][ii]]!=1)
    {
      move(ii-1);
    }
  }
  while(ant[a-1][1]!=base)
{ ant[a-1][0]=ant[a][0];
    ant[a-1][1]=ant[a][1];
    ant[a][0]=0;
      ant[a][1]=0;
      a--;
  }
}
//------------------------------------------------------
bool visit(int node,int c){

int sw=0;
for(int ii=0;ii<kn;ii++)
{
   if(path[c][ii]==node)
   sw=1;

}
if(sw==0)
return true;
else
return false;
}
//------------------------------------------------------
void update_path(int ii){   //Update backtracking paths
    int jj=0;
```

```
 while((path[num_paths-1][jj]!=ii)&&(jj<MAX_NODES))
 {path[num_paths][jj]=path[num_paths-1][jj];
  jj++; kk++;
  }


 path[num_paths][jj]=path[num_paths-1][jj];

}
//----------------------------------------------------------
//----------------------------------------------------------
int Count(int i,int num_paths){ //Count nodes in a path
int cont=0;
int np=0;
np=num_paths;
while(path[np][cont]!=i)
{
   cont++;
}
return cont+1;
}
//----------------------------------------------------------
//backtracking search paths----------------------------
void Search_Paths(int i)
{
         if(i<n)
         {
                 for(int j=ind;j<=n;j++)
                 {
                  if(visited[i][j]==1)
                  {

                  if(kk==1)
                 kn=Count(i,num_paths-1);
                 else
                 kn=Count(i,num_paths);

                 int np=0;

                  if(kk>1)
                 np=num_paths;
```

```
                    else
                np=num_paths-1;
                if(visit(j,np))
                    {
                        if((kk==1)&&(num_paths>0))
                        update_path(i);

                    path[num_paths][kk]=j;

                    if(j!=target)
                        {
                            ant[a][0]=i;
                            ant[a][1]=j;
                            a++;
                            kk++;
                            ind=0;
                            Search_Paths(j);
                        }else{
                            path[num_paths][kk+1]=-1;
                            num_population_initial++;
                            num_paths++;
                            ant[a][0]=i;
                            ant[a][1]=j;
                            a++;
                            i=ant[a-1][0];
                            ind=ant[a-1][1]+1;
                            j=ind;
                            ant[a][0]=0;
                            ant[a][1]=0;
                            a--;
                            del_visited(i);
                            update_population();
                            kk=1;
                        }//end else
                    }//end visit
                }//end population
            }//end for
        }
}
//End backtracking search paths-----------------------
```

```
//Remove paths function 2-------------------------------
void remove_path_2(int i_del){
        for(int i=i_del;i<num_population-1;i++){
            int j=0;
            while(population[i+1][j]!=-1)
              {
            population[i][j]=population[i+1][j];
            j++;
            }
                for(int k=j;k<MAX_NODES;k++)
                {population[i][k]=-1;}
        }
      num_population--;
      int n=population[0][0];
}
//-----------------------------------------------------
//Delete repeated paths------------------------------
void repetition(int p[MAX_POPULATION][MAX_POPULATION],int
ii,int n){

int v=0;

        for(int i=ii+1;i<num_population;i++){
            int j=0;v=0;
            while (population[i][j] != -1) {

              if(population[i][j]==p[ii][j])
              v++;
              else
              v=0;

              j++;
            }

            if(v==j)
            {remove_path_2(i);
              i--;
              }
        }
  }
```

```
//---------------------------------------------------------
//Path Feasibility analysis-------------------------------
void feasibility(int j){

int fit=0;
int k=0;
while (population[j][k+1] != -1) {

 fit = fit + capacity[population[j][k]]
 [population [j][k+1]];  //capacity
 k++;
 }
if(fit<demand){
    remove_path_2(j);
}
}
//---------------------------------------------------------
//initial population--------------------------------------
void initial_population() {

   int i,j,i1,j1,j2;
   int num_paths_by_destination_tmp[MAX_NODES][2];

   for (i=0; i<MAX_POPULATION; i++) {
    for (j=0; j<MAX_POPULATION; j++) {
        new_population[i][j] = -1;
        new_population2[i][j] = -1;
        new_population3[i][j] = -1;
     }
    }

   for(int ii=0;ii<=num_population;ii++)
   repetition(population,ii,num_population);

for(int i=0;i<num_population;i++)
feasibility(i);

i=0;
   while(population[i][0]!=-1 && population[i][1]!=0){
   i++;
   }
```

```
 num_population=i;
 num_population_initial=i;
}
//end initial_population()------------------------------
//SetFit()---------------------------------------------
void SetFit(int p,int i, int s){

    strength[i][p]=s;
}
//end SetFit()-----------------------------------------
//GetFit()---------------------------------------------
float GetFit(int p,int i){

    return strength[i][p];
}
//end GetFit()-----------------------------------------
//dominatesOrEqual-------------------------------------
bool dominatesOrEqual(int i,int j,int
cam1[MAX_NODES][MAX_NODES],int
cam2[MAX_NODES][MAX_NODES]){

    int k=0,l,state1,state2;
    int win,loss,tie;

 int fit[2][2];
 fit[0][0]=0;
 fit[0][1]=0;
 fit[1][0]=0;
 fit[1][1]=0;

 while (cam1[i][k+1] != -1) {
 fit[0][0] = fit[0][0]+1;//hop count
 fit[0][1] = fit[0][1] + delay[cam1[i][k]][cam1[i][k+1]];
     //delay
     k++;
     }
     k=0;

 while (cam2[j][k+1] != -1) {
 fit[1][0] = fit[1][0]+1;//hop count
     fit[1][1] = fit[1][1] +
```

```
            delay[cam2[j][k]][cam2[j][k+1]];//delay
            k++;
            }
                win    = 0;
                loss = 0;
                tie = 0;
                if (fit[0][0] < fit[1][0])
                    win++;
                else if (fit[0][0] == fit[1][0])
                        tie++;
                  else
                        loss++;


                if (fit[0][1] < fit[1][1])
                    win++;
                else if (fit[0][1] == fit[1][1])
                        tie++;
                  else
                        loss++;


        if(win>loss)
        return true;
        else
            if(win==loss)
            return true;
            else
            return false;


}
//end dominatesOrEqual--------------------------------
//Calculate Non-Dominated Fitness--------------------
void  calcStrengths()
{

    int   i, j;
    int   domCount,pop_count;

    for (i = new_num_population1 - 1; i >= 0; i--) {

        domCount = 0;
```

```
    pop_count=0;
    for (j = num_population - 1; j >= 0; j--){
      if(population[j][0]!=-1){
      pop_count++;
      if (dominatesOrEqual(i,j,new_population1,
        population)){
          domCount++;
      }
      }
    }
    SetFit(0,i, 1/(double(domCount) / double(pop_count +
1))); //num_population
  }
}
//end calcStrengths()-------------------------------------
//Calculate dominated fitness----------------------------
void  calcFitness()
{
  int     i, j;
  double  fitness1;
  for (i = num_population - 1; i >= 0; i--) {
    fitness1 = 0.0;
   if(population[i][0]!=-1){
   for (j = new_num_population1 - 1; j >= 0; j--) {
    if (dominatesOrEqual(j,i,new_population1,population))
        fitness1 += GetFit(0,j);
   }
    SetFit(1,i, 1.0 + fitness1);
    }
  }
}
//end calcFitness()--------------------------------------
//Distance calculate-------------------------------------
double  IndDistance(int i,int j,int
cam1[MAX_NODES][MAX_NODES])
{
double distance;
distance=0;

        int fit[2][2],k=0;
```

```
                fit[0][0]=0;
                fit[0][1]=0;
                fit[1][0]=0;
                fit[1][1]=0;

                while (cam1[i][k+1] != -1) {
                fit[0][0] = fit[0][0]+1;//hop count
                    fit[0][1] = fit[0][1] +
                    delay[cam1[i][k]][cam1[i][k+1]];//delay
                    k++;
                    }
                    k=0;

                while (cam1[j][k+1] != -1) {
                fit[1][0] = fit[1][0]+1;//hop count
                    fit[1][1] = fit[1][1] +
                    delay[cam1[j][k]][cam1[j][k+1]];//delay
                    k++;
                    }
            distance = sqrt(
                        (pow(fit[0][0] - fit[1][0], 2)) +
                        (pow(fit[0][1] - fit[1][1], 2))
                         );
        return distance;
    }
    //End distance------------------------------------------
    //Cluster distance------------------------------------------
    double  ClusterDistance(int   ic1, int   ic2)
    {
        double  sum = 0;
        int   numOfPairs = 0;
        int ik1=0;
        while(clusters[ic1][ik1]!=-1)
        {
                int ik2=0;
                while(clusters[ic2][ik2]!=-1){
                  numOfPairs++;
                  sum += IndDistance(ic1,ic2,new_population);
                   ik2++;
                }
```

```
            ik1++;
    }

if(sum<0)  sum=sum*-1;

return (sum / double(numOfPairs));

}
//End clusters distance--------------------------------
//Join Clusters----------------------------------------
void  JoinClusters(int  cluster1, int  cluster2, int&
numOfClusters)
{

int  k=0;
while(clusters[cluster1][k]!=-1)
{
  k++;
}

int  kk=0;
while(clusters[cluster2][kk]!=-1)
{clusters[cluster1][k]=clusters[cluster2][kk];
k++;kk++;
}

k=0;
while(clusters[cluster2][k]!=-1)
{clusters[cluster2][k]=-1;k++;}

numOfClusters--;
}
//End Join Clusters------------------------------------
//Find Cluster Centroid--------------------------------
void  ClusterCentroid(int cluster)
{

          int  cluster_value[200][2],k=0,prom1=0,prom2=0;

          for(int  i=0;i<100;i++)
```

```
      {cluster_value[i][0]=0;cluster_value[i][1]=0;}

      int num_cam=0;
      while(clusters[cluster][k]!=-1){
            int cl=clusters[cluster][k];
            int kk=0;
              while (new_population[cl][kk+1] != -1) {
                cluster_value[cluster][0] =
cluster_value[cluster][0]+1;//hop count
            cluster_value[cluster][1] = cluster_value
[cluster][1]+delay[new_population[cl][kk]][new_population
[cl][kk+1]];//delay
          kk++;
          }

        num_cam++;
        prom1=cluster_value[cluster][0];
        prom2=cluster_value[cluster][1];
        k++;

      }
      prom1=prom1/num_cam;
      prom2=prom2/num_cam;

  for(int i=0;i<100;i++)
  {cluster_value[i][0]=0;cluster_value[i][1]=0;}
  int min_dist=1000,selected=0;
  float dist=0;
  k=0;

  while(clusters[cluster][k]!=-1){
  kk=0;
  int cl=clusters[cluster][k];
    cluster_value[cluster][0]=0;
    cluster_value[cluster][1]=0;
    while (new_population[cl][kk+1] != -1) {
        cluster_value[cluster][0] = cluster_value
        [cluster][0]+1;
      //hop count
        cluster_value[cluster][1] = cluster_value
```

```
               [cluster][1]  +
delay[new_population[cl][kk]][new_population[cl][kk+1]];
//delay
           kk++;
    }
       dist=sqrt(pow(cluster_value[cluster][0]-
prom1,2)+pow(cluster_value[cluster][1]-prom2,2));
       if(dist<min_dist)
       {
           selected=clusters[cluster][k];
           min_dist=dist;
       }
     k++;

   }
    clusters[cluster][0]=selected;
    k=1;
    while(clusters[cluster][k]!=-1){
        clusters[cluster][k]=-1;
        k++;
   }
   k=0;
   while(new_population[clusters[cluster][0]][k]!=-1){
new_population1[new_num_population1][k]=new_population
[clusters[cluster][0]][k];
   k++;
   }
   new_population1[new_num_population1][k]=-1;

   new_num_population1++;

}
//End Cluster Centroid-----------------------------------
//Reduce Non-Dominated-----------------------------------
void   clustering(void)
{
int i;
for (i = 0 ; i < new_num_population; i++){
clusters[i][0]=i;
}
```

```
    int population_clusters=new_num_population;
    while(population_clusters>maxNondominated){

        float minDist=100;
        int join1,join2;
        int   numOfClusters = new_num_population;
        for (int   c = 0; c < numOfClusters; c++) {
        for (int   d = c + 1; d < numOfClusters; d++) {
          if(clusters[c][0]!=-1 && clusters[d][0]!=-1)
          {
                    double dist=ClusterDistance(c,d);
//clusters[c][0],clusters[d][0]);

          if (dist < minDist) {
            minDist = dist;
             join1 = c;
             join2 = d;
                  }}
        }
        }

    JoinClusters(join1, join2, population_clusters);
    }
      int c=0,ind=0;
      while(c<population_clusters)
      {
      if(clusters[ind][0]!=-1)
      {
      ClusterCentroid(ind);
      c++;
      }
      ind++;
      }
}
//End Reduce Non-Dominated------------------------------
//Bubble_Sort------------------------------------------
void Bubble_Sort()
{
   int   i,j;
   float temp1=0,temp2=0;
```

```
      for(i=0;i<new_num_population1;i++){
          for(j=i+1;j<new_num_population1;j++){

                  if(Prob[i][0]>Prob[j][0])
                  {
                     temp1=Prob[i][0];
                     temp2=Prob[i][1];
                     Prob[i][0]=Prob[j][0];
                     Prob[i][1]=Prob[j][1];
                     Prob[j][0]=temp1;
                     Prob[j][1]=temp2;
                  }
          }
      }

      for(i=0;i<num_prob;i++){
          for(j=i+1;j<num_prob;j++){

                  if(Prob[i][2]>Prob[j][2])
                  {
                     temp1=Prob[i][2];
                     temp2=Prob[i][3];
                     Prob[i][2]=Prob[j][2];
                     Prob[i][3]=Prob[j][3];
                     Prob[j][2]=temp1;
                     Prob[j][3]=temp2;
                  }
          }
      }
}
//End Bubble_Sort-----------------------------------------
//Join_Prob----------------------------------------------
void Join_Prob()
{
   int  temp1=0,temp2=0,i,j,ii=0;

   Jprob[0][0]=0;
   Ci[0][7]=0;
   Ci[0][8]=0;
```

```
        for(i=0;i<num_prob;i++){

            if(i==0){
            Jprob[i][0]=Prob[i][2];
            }else{
            Jprob[i][0]=Jprob[i-1][0]+Prob[i][2];
            }
            Jprob[i][1]=1;
            Jprob[i][2]=Prob[i][3];
        }
        j=i;
        for(i=0;i<new_num_population1;i++){
        Jprob[j][0]=Jprob[j-1][0]+Prob[i][0];
        Jprob[j][1]=0;
        Jprob[j][2]=Prob[i][1];
        j++;
        }
}
//End Join_Prob-------------------------------------------
//Set_new_population---------------------------------------
void Set_new_population(int p,int i)
{
int k;

if(p==0){
 k=0;
   while(new_population1[i][k]!=-1)
   {

new_population[new_num_population][k]=new_population1[i][
k];
     k++;
   }
   new_population[new_num_population][k]=-1;
   new_num_population++;
}else{
k=0;
   while(population[i][k]!=-1)
   {
   new_population[new_num_population][k]=population[i][k];
```

```
  k++;
  }
  new_population[new_num_population][k]=-1;
  new_num_population++;
}
}
//Set_new_population---------------------------------------
//Roulette_Selection---------------------------------------
void  Roulette_Selection()
{
    float num_random,sum_fitness=0,sum_ci=0,prob[100][3];
    int i,j;
    for(i=0;i<MAX_NODES;i++)
    {   for(j=0;j<MAX_NODES;j++)
        {
          new_population[i][j]=-1;
          }
        }

  new_num_population=0;

  for(i=0;i<new_num_population1;i++){
  sum_fitness=sum_fitness+GetFit(0,i);
  Ci[i][0]=0;
  }

  for(i=0;i<num_population;i++){
  if(population[i][0]!=-1){
  sum_fitness=sum_fitness+GetFit(1,i);
  Ci[i][1]=0;
  }
  }

  for(i=0;i<new_num_population1;i++){
  Ci[i][0]=1-double(GetFit(0,i)/sum_fitness);
  sum_ci=sum_ci+Ci[i][0];
  }

  for(i=0;i<num_population;i++){
  if(population[i][0]!=-1){
```

```
Ci[i][1]=1-double(GetFit(1,i)/sum_fitness);
sum_ci=sum_ci+Ci[i][1];
}
}

for(i=0;i<new_num_population1;i++){
Prob[i][0]=Ci[i][0]/sum_ci;
Prob[i][1]=i;
}
j=0;
for(i=0;i<num_population;i++){
if(population[i][0]!=-1){
Prob[j][2]=Ci[i][1]/sum_ci;
Prob[j][3]=i;
j++;
}
}

num_prob=j;
Bubble_Sort();
Join_Prob();

new_num_population=0;
for(i=0;i<(num_prob+new_num_population1)/2;i++){
num_random=rand()/32300.0;
int sw=0;
j=0;
while((j<num_population+new_num_population1-
1)&&(sw==0)){
if(Jprob[0][0]>num_random){
Set_new_population(Jprob[0][1],Jprob[0][2]);
sw=1;
}else{

if((Jprob[j+1][0]>=num_random)&&(Jprob[j][0]<num_random))
{
Set_new_population(Jprob[j+1][1],Jprob[j+1][2]);
sw=1;
}
j++;
```

```
        }
        }
        }
}
//end Roulette_Selection---------------------------------
void select_non_dominated()   {

  for (int i=0; i<MAX_POPULATION; i++) {
     for (int j=0; j<MAX_POPULATION; j++) {
         new_population[i][j] = -1;
     }
        }

        int  i,j,sw,k,l,state1,state2;
  int win,loss,tie;
new_num_population = 0;

  for(i=0; i<num_population; i++) {
                j=0;
                sw=0;
                if(population[i][0]!=-1)
                {
                    while((j<num_population)&&(sw==0)){
                    win    = 0;
     loss = 0;
     tie = 0;

                    if (fitness[i][0] < fitness[j][0])
        win++;
     else if (fitness[i][0] == fitness[j][0])
            tie++;
          else
             loss++;

     if (fitness[i][1] < fitness[j][1])
         win++;
     else if (fitness[i][1] == fitness[j][1])
            tie++;
          else
             loss++;
```

```
                    if (win > 0) {
        if (loss == 0) {
            state1 = NONDOMINATED;
            state2 = DOMINATED;
        }
        else {
            state1 = BOTH_NONDOMINATED;
            state2 = BOTH_NONDOMINATED;
        }
    }
    else {
        if (loss > 0) {
                                        sw=1;
            state1 = DOMINATED;
            state2 = NONDOMINATED;
        }
        else {
            state1 = BOTH_NONDOMINATED;
            state2 = BOTH_NONDOMINATED;
        }
    }

    if (tie == 4) {
        state1 = NONDOMINATED;
        state2 = DOMINATED;
    }
                    j++;
                }
                }
                        if(sw==0){
                            k=0;
                    while (population[i][k] != -1) {
                new_population[new_num_population][k] =
population[i][k];
                            population[i][k]=-1;
            k++;
        }
        new_population[new_num_population][k] =
population[i][k];
```

```
                                           population[i][k]=-1;
                k++;
            new_num_population++;
                                 }
    }

            for(i=0; i<num_population; i++){
                    if(population[i][k]==-1)
                    {
                    j=i;
                            while(j<num_population){
                                    k=0;
                            while(population[j+1][k]!=-1){
                        population[j][k]=population[j+1][k];
                                    k++;
                                    }
                                    population[j][k]=-1;
                            j++;
                            }
                            num_population--;
                    }
            }
    if (num_population == 1) {
        k=0;
        for (l=0; l < d; l++) {
            while (population[0][k] != -1) {
                new_population[0][k] = population[0][k];
                                 population[i][k]=-1;
                k++;
            }
            new_population[0][k] = population[0][k];
                             population[i][k]=-1;
            k++;
        }
        new_num_population++;
    }
}
//SPEA------------------------------------------------------
void SPEA()
{
```

```
int i,k;
  select_non_dominated();
  if (new_num_population > maxNondominated){
  new_num_population1=0;
  clustering();
  }else{
  new_num_population1=0;
  for(i=0;i<new_num_population;i++){
        k=0;
        while(new_population[i][k]!=-1){

new_population1[new_num_population1][k]=new_population[i]
[k];
        k++;
        }
        new_population1[new_num_population1][k]=-1;
        new_num_population1++;
  }
  }
  calcStrengths();
  calcFitness();
}
//End SPEA--------------------------------------------------
//Functions Calculate --------------------------------------
void function_calculate(int
pop[MAX_POPULATION][MAX_NODES], int num_pop) {

  int i,j,k,max,min,max2,min2;
        max=0;
        min=1000;
        max2=0;
        min2=1000;
  float alpha;

  for (i=0; i<num_pop; i++) {
      for (j=0; j<2; j++) {
          fitness[i][j] = 0.0;
      }
  }
  for (i=0; i<num_pop; i++) {
```

```
    k=0;
    for (j=0; j < d; j++) {
        while (pop[i][k+1] != -1) {
            fitness[i][0] = fitness[i][0] + 1;//hop
count
            fitness[i][1] = fitness[i][1] +
delay[pop[i][k]][pop[i][k+1]];//delay
            k++;
        }


    }
 }


}
//end functions calculate-------------------------------
//remove path-----------------------------------------
void remove_path(int i_del,int paths-tmp
[MAX_NODES][MAX_NODES],int num){

for(int k=0;k<i_del;k++)
{
    for(int j=0;j<num;j++){
     paths-tmp[del_path[k]][j]=0;
     }
        for(int i=del_path[k];i<num-1;i++){
            int j=0;
            while(paths-tmp[i][j]!=-1)
             {
            paths-tmp[i][j]=paths-tmp[i+1][j];
            j++;
            }
        }
        num--;
}
}
//-----------------------------------------------------
//end remove path--------------------------------------
//Constraint-------------------------------------------
void restriction(int paths-tmp[MAX_NODES][MAX_NODES],int
num){
```

```
for(int i=0;i<num;i++){
        int j=0,sw=0;
        while((paths-tmp[i][j]!=-1)&&(sw==0)){
                if(capacity[paths-tmp[i][j]][paths-
tmp[i][j+1]]<demand)
                {
                if(capacity[paths-tmp[i][j]][paths-
tmp[i][j+1]]!=0){


                    for(int ii=i;ii<=num;ii++){
                      int jj=0;

                        while(paths-tmp[ii][jj]!=-1)
                        {
                        paths-tmp[ii][jj]=paths-tmp[ii+1][jj];
                          jj++;


                        }
                    }
                sw=1;
                }
              }
              j++; }
        }
}
//end constraint------------------------------------------
//cross-over pointer-------------------------------------
int crossover_position(int i){ //Obtain the position-Cut
of a path
int sw=0,j=0;
while(sw==0){
        if((new_population[i][j]==-
1)||(new_population[i+1][j]==-1)){
                sw=1;
        }
        j++;
}
return j-2;
}
//end cross-over pointer--------------------------------
```

```
//verify_path---------------------------------------------
bool verify_path(int pop[MAX_NODES][MAX_NODES], int
indice){

    int j=0,sw=0;
              while(pop[indice][j+1]!=-1)
              {
                    if(capac-
ity[pop[indice][j]][pop[indice][j+1]]==0)
                    sw=1;
              j++;
              }

    if(sw==0)
    return false;
    else
    return true;


}
//end verify_path----------------------------------------
//remove_path_3-------------------------------------------
void remove_path_3(int indice, int
pop[MAX_NODES][MAX_NODES],int size){

      int i=0,j;

        for(i=indice;i<MAX_NODES;i++)
        pop[indice][i]=-1;

        for(i=indice;i<size-1;i++){
            j=0;
            while(pop[i+1][j]!=-1)
             {
            pop[i][j]=pop[i+1][j];
            j++;
            }

          for(int k=j;k<MAX_NODES;k++){pop[i][k]=-1;}
        }
}
```

```
//end remove path----------------------------------------
//Repeated_nodes ----------------------------------------
bool Repeated_nodes(int pop[MAX_NODES][MAX_NODES], int
indice){
int i=0,j=0,sw=0;

while(pop[indice][i+1]!=-1)
{
    j=0;
    while(pop[indice][j]!=-1){

            if(i!=j){

                    if(pop[indice][i]==pop[indice][j]){
                       sw=1;
                    }
            }
             j++;
            }
        i++;
}
if(sw==0)
return false;
else
return true;
}
//End Repeated_nodes ----------------------------------
//cross over ----------------------------------------
void cross_over() {

for (int i=0; i<MAX_POPULATION; i++) {
    for (int j=0; j<MAX_POPULATION; j++) {
        new_population2[i][j] = -1;
    }
  }

  int i,l,k,k1,k2,k3,cross_over_position;
  float num_random;
  new_num_population2 = 0;
```

```
if(new_num_population == 1){
        k3=0;
        while(new_population[0][k3]!= -1){
    new_population2[0][k3]=new_population[0][k3];
        k3++;
        }
    new_population2[0][k3]=new_population[0][k3];

}else{

    for(i=0; i<new_num_population-1; i++){
    num_random = rand()/32300.0;   //Evaluate CrossOver
Prob

    if (num_random <= CROSS_OVER_PROBABILITY) {

            cross_over_position=ran-
dom(crossover_position(i));
            if(cross_over_position==0){
            cross_over_position=1;
            }

            k=0;k1=0;k2=0;
            while(k<cross_over_position){
            new_population2[new_num_population2][k]
= new_population[i][k];
            new_population2[new_num_population2+1][k]
= new_population[i+1][k];
            k++;
            }
            k1=k;
            k2=k;
            while(new_population[i+1][k1]!= -1){
new_population2[new_num_population2][k1]=new_population[i
+1][k1];
            k1++;
            }
new_population2[new_num_population2][k1]=new_population[i
+1][k1];
```

```
                         while(new_population[i][k2]!= -1){

new_population2[new_num_population2+1][k2]=new_population
[i][k2];
                         k2++;
                         }

new_population2[new_num_population2+1][k2]=new_population
[i][k2];
                new_num_population2=new_num_population2+2;
                }
                }
                }

                for(i=0;  i<new_num_population;  i++){

                         int k4=0;
                         while(new_population[i][k4]!= -1){

new_population2[new_num_population2][k4]=new_population[i
][k4];
                         k4++;
                         }

new_population2[new_num_population2][k4]=new_population[i
][k4];
                         k4++;
                         new_num_population2++;
                }


        for(k=0;k<new_num_population2;k++)
        {
        if(Repeated_nodes(new_population2,k))
      {remove_path_3(k,new_population2,new_num_population2);
         new_num_population2--;
         k--;
         }else{
                if(verify_path(new_population2,k))

{remove_path_3(k,new_population2,new_num_population2);
```

```
                  new_num_population2--;
                  k--;
                }
          }
        }
}
//end cross_over()----------------------------------------
//mutation------------------------------------------------
int m_position(int i){ //Obtain the position-Cut of a path
int sw=0,j=0;
while(sw==0){
          if(new_population[i][j]==-1){
                sw=1;
          }
          j++;
}
return j-2;
}
//end_mutation_position-----------------------------------
//mutation------------------------------------------------
void mutation() {

for (int i=0; i<MAX_POPULATION; i++) {
   for (int j=0; j<MAX_POPULATION; j++) {
       new_population3[i][j] = -1;
   }
  }

 int i,l,k,k1,k2,u,i2,mutation_position,p;
 float num_random;
         int paths_tmp[MAX_NODES];

 new_num_population3 = 0;

         for(i=0; i<new_num_population2-1; i++){
         num_random = rand()/32300.0;

             if (num_random <= MUTATION_PROBABILITY){

                mutation_position=random(m_position(i));
```

```
            if(mutation_position==0){
            mutation_position=1;
            }
            k=0;k1=0;

                    while(k<mutation_position){
            new_population3[new_num_population3][k]
= new_population2[i][k];
                    k++;
                    }

                    if(bfs(new_population2[i][k-
1],dest[0])){

                        paths_tmp[0] = dest[0];
        for (u=dest[0]; pred[u]>=0; u=pred[u]) {
                            k1++;
            paths_tmp[k1] = pred[u];
                        visited[pred[u]][u] = 0;
        }


        new_population3[new_num_population3][k] =
pred[u];

        for (int i1 = k1-1; i1>=0; i1--) {
new_population3[new_num_population3][k]=paths_tmp[i1];
        k++;
        }
                        new_num_population3++;
                        }
                }
            }


    for(k=0;k<new_num_population3;k++)
    {
    if(Repeated_nodes(new_population3,k))
    {remove_path_3(k,new_population3,new_num_population3);
        new_num_population3--;
```

```
       k--;
     }else{
          if(verify_path(new_population3,k))

{remove_path_3(k,new_population3,new_num_population3);
              new_num_population3--;
              k--;
              }
       }
     }
}
//end mutation()-----------------------------------------
bool rest(int p[MAX_POPULATION][MAX_POPULATION],int i)
{        bool val=true;
         int j=0;
         while (p[i][j] != -1) {
              if(capacity[p[i][j]][p[i][j+1]]<demand)
              {
                if(capacity[p[i][j]][p[i][j+1]]>0)
              {
                 val=false;
                 }
              }
            j++;
          }
         return val;
}
//------------------------------------------------------
void Clear_cluster(){

   for(int i=0;i<100;i++){
   for(int j=0;j<500;j++){
   clusters[i][j]=-1;
   }
   }
}
//UpDate new populations--------------------------------
void update_new_population() {
 int i,j,k,l;
         int num_temp_population=0;
```

```
             for (i=0; i<MAX_POPULATION; i++) {
      for (j=0; j<MAX_POPULATION; j++) {
                    temp_population[i][j] = -1;
                    }
  }
          int ii=0;
  for (i=0; i<num_population; i++) {
          j=0;
                  if(population[i][0]!=-1){
      while(population[i][j]!=-1){
                  temp_population[ii][j]=population[i][j];
                       j++;
                  }
                  temp_population[ii][j]=-1;
                  ii++;
                  }
  }

          num_temp_population=ii;

          for (i=0; i<MAX_POPULATION; i++) {
         for (j=0; j<MAX_POPULATION; j++) {population[i][j]=
-1;}}

          num_population = 0;
          for(i=0;i<new_num_population1;i++){
          j=0;
            if(new_population1[i][j]!=-1)
            {
      while(new_population1[i][j]!=-1){
                    popula-
tion[num_population][j]=new_population1[i][j];
                    j++;
                  }
                  population[num_population][j]=-1;
                  num_population++;
            }
            }
        for(i=0; i<new_num_population2; i++) {
      k=0;
      while (new_population2[i][k] != -1) {
```

```
     population[num_population][k] =
new_population2[i][k];
     k++;
     }
     population[num_population][k] =
new_population2[i][k];
     k++;
                    num_population++;
 }

 for(i=0; i<new_num_population3; i++) {
     k=0;
     while (new_population3[i][k] != -1) {
     population[num_population][k] =
new_population3[i][k];
     k++;
     }
     population[num_population][k] =
new_population3[i][k];
     k++;
                    num_population++;

 }
        for (i=0; i<MAX_POPULATION; i++) {
     for (j=0; j<MAX_POPULATION; j++) {
                    new_population[i][j] = -1;
                    new_population1[i][j] = -1;
                    new_population2[i][j] = -1;
                    new_population3[i][j] = -1;
                    temp_population[i][j] = -1;
                    }
 }

     Clear_cluster();
}
//end update_new_population()-----------------------------
//population_view()-----------------------------------
void population_view(int
population_v[MAX_POPULATION][MAX_POPULATION], int
```

```
size_population,int count_generations ,String title) {
//char *title
 int i,k,l,g,h;

 printf("\n\n%s\n\n",title);
         printf("GENERATION \tPOPULATION\n");
 for(i=0; i < size_population; i++) {
     k=0;
                     printf("%d \t\t",count_generations);
         while (population_v[i][k] != -1) {
             printf("%d,",population_v[i][k]);
                                 k++;

         }
                             printf("\n");
         k++;

 }
}
//end population_view()----------------------------------
//FITNESS VIEW------------------------------------------
void fitness_view(int generation, int tam) {
 int i;

 printf("\FUNCTION VALUES\n\n");
         printf("CHROMOSOME \t FUNCTIONS \t GENERATION\n");
 for(i=0; i < tam; i++) {
     printf("Chromosome %d \t Hop Count \t %f\n",i,
fitness[i][0]);
     printf("Chromosome %d \t Delay      \t %f\n",i,
fitness[i][1]);
         }
 printf("\n");
}
//end fitness_view()------------------------------------
//------------------------------------------------------
void B_INPUTFILE()
{
int a,b,c,h,i,j;
     FILE* input = fopen("mf_MOEA.in","r");
     // read number of nodes, edges and sinks
     fscanf(input,"%d %d %d\n",&n,&e,&d);
```

```
    // initialize empty capacity matrix
    for (i=0; i<n; i++) {
    num_paths_by_destination[i][0] = 0;
    num_paths_by_destination[i][1] = 0;
for (j=0; j<n; j++) {
    capacity[i][j] = 0;
    delay[i][j] = 0;
    visited[i][j] = 0;
}

    }
    for (i=0; i<MAX_POPULATION; i++) {
for (j=0; j<MAX_POPULATION; j++) {
    population[i][j] = 0;
}

    }

    // read sink nodes
printf("SINK NODE: ");
for (i=0; i<d; i++) {
    fscanf(input,"%d",&dest[i]);
    printf("%d\n",dest[i]);

                    target=dest[i];
}
// read demand of the multicast flow (kbps)
printf("DEMAND FLOW(kbps): ");
fscanf(input,"%d\n",&demand);

        printf("%d\n\n",demand);
    // read edge capacities
printf("Edge Capacities \n");

    for (i=0; i<e; i++) {
    fscanf(input,"%d %d %d %d",&a,&b,&c,&h);
     printf("%d %d %d %d\n",a,b,c,h);

                capacity[a][b] = c;
    delay[a][b] = h;
    visited[a][b] = 1;
    }
```

```
    int  r=0;
    r=visited[8][12];

    fclose(input);
}
//------------------------------------------------------------
void OUTFILE(int size_population,int
population_v[MAX_POPULATION][MAX_POPULATION]){
FILE* output = fopen("OUTPUT_FILE.txt","w");
    // read number of nodes, edges and sinks
    fprintf(output,"%s\n\n","FINAL POPULATION");
    fprintf(output,"GENERATION \tPOPULATION\n\n");
        int i,k;
  for(i=0; i < size_population; i++) {
    k=0;
                                fprintf(output,"%d \t\
t",count_generations);
        while (population_v[i][k] != -1) {
            fprintf(output,"%d,",population_v[i][k]);
                                    k++;
        }
                                fprintf(output,"\n");
        k++;
  }

        fprintf(output,"\nFUNCTION VALUES\n\n");
        fprintf(output,"CHROMOSOME \t FUNCTIONS \t
GENERATION\n\n");
        for(i=0; i < size_population; i++) {
      fprintf(output,"Chromosome %d \t Hop Count\t %f\
n",i,fitness[i][0]);
      fprintf(output,"Chromosome %d \t Delay    \t %f\
n",i,fitness[i][1]);
        }
  fprintf(output,"\n");
        fclose(output);
}
//------------------------------------------------------------
int main ()
{
```

```
      B_INPUTFILE();
      maxNondominated = MAX_NONDOMINATED_SOLUTIONS;
      int   nondominatedSetUpperBound = num_population
+ maxNondominated;

      time1 = clock();
      Search_Paths(0);
 initial_population();

     population_view(population,num_population_initial
,0,"INITIAL POPULATION");
         num_population=num_population_initial   ;
 count_generations = 0;

 while (count_generations < MAX_GENERATIONS) {
     function_calculate(population,num_population);
     Clear_cluster();
                 SPEA();
                 Roulette_Selection();
     cross_over();

population_view(new_population2,new_num_population2,count
_generations,"POPULATION AFTER CROSSOVER");
     mutation();

population_view(new_population3,new_num_population3,count
_generations,"POPULATION AFTER MUTATION   ");
     update_new_population();
     count_generations++;
 }

       for(int  ii=0;ii<num_population;ii++)
       repetition(population,ii,num_population);

       function_calculate(population,num_population);
       select_non_dominated();

population_view(new_population,new_num_population,4,"FINAL
POPULATION");
function_calculate(new_population,new_num_population);
```

```
        fitness_view(count_generations,new_num_population);
            time2 = clock();
 printf("\nMOEA running time : %f \n",(float)(time2-
time1)/CLK_TCK*1000);
            OUTFILE(new_num_population,new_population);
 getch();
      return 0;
}
//------------------------------------------------------------
```

Input File (mf_MOEA.in)

```
14 40 1        /* nodes Links Number of Destinations */
12             /* Destination Node */
256            /* Demand of the flow (Kbps) */
0 1 1536 9     /* Node i Node j Link Capacity Delay */
0 2 1536 9
0 3 1536 7
1 0 1536 9
2 0 1536 9
3 0 1536 7
1 3 1536 13
1 6 1536 20
3 1 1536 13
6 1 1536 20
2 4 1536 7
2 7 1536 26
4 2 1536 7
7 2 1536 26
3 10 1536 15
10 3 1536 15
4 5 1536 7
4 10 1536 11
5 4 1536 7
10 4 1536 11
5 6 1536 7
6 5 1536 7
6 9 1536 7
9 6 1536 7
7 8 1536 5
```

```
8  7  1536  5
7  13  1536  8
13  7  1536  8
8  9  1536  5
9  8  1536  5
8  12  1536  7
12  8  1536  7
9  11  1536  8
11  9  1536  8
10  12  1536  14
12  10  1536  14
10  11  1536  9
11  10  1536  9
12  13  1536  4
13  12  1536  4
```

Example 1

Destination node: 12

SOURCE	0
TARGET	12

		FUNCTIONS	
SOL	CHROMOSOME	HOP	DELAY
1	0,3,10,12	3	36

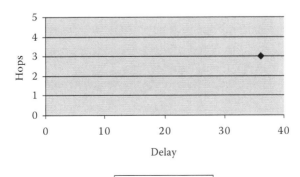

◆ Opt. Pareto Front

Topology

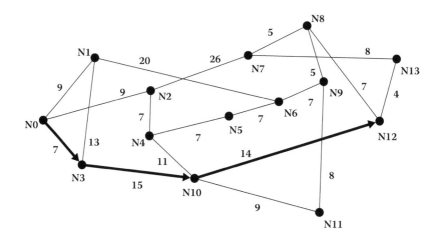

Example 2

Destination node: 13

SOURCE	0
TARGET	13

SOL	CHROMOSOME	FUNCTIONS	
		HOP	DELAY
1	0,2,7,13	3	43
2	0,3,10,12,13	4	40

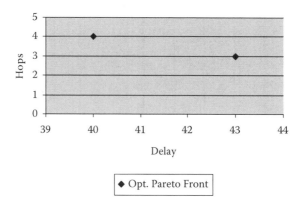

♦ Opt. Pareto Front

Topology

First solution:

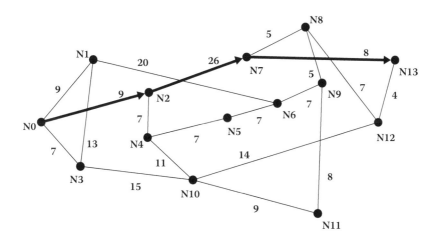

Second solution:

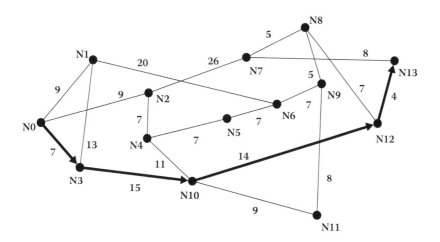

Program in C to Section 4.7.2

```
//-----------------------------------------------------
/*
Program: MOEA - MULTICAST
*/
```

```
#include <vcl.h>
#pragma hdrstop

#include <stdlib.h>
#include <math.h>
#include <stdio.h>
#include <conio.h>
#include <time.h>
//------------------------------------------------------
---------------------
//Basic Definitions
#define WHITE 0
#define GRAY 1
#define BLACK 2
#define MAX_NODES 50
#define MAX_POPULATION 50
#define MAX_GENERATIONS 5
#define SIZE_COMPARISON 1.0
#define CROSS_OVER_PROBABILITY 0.7
#define MUTATION_PROBABILITY 0.3
#define ERROR 0.00001
#define oo 1000000000
#define oof 10000.0
#define zero 0.000001
#define DOMINATED 0
#define NONDOMINATED 1
#define BOTH_NONDOMINATED 2
#define MAX_NONDOMINATED_SOLUTIONS 3
//

//Declarations
float Prob[100][4];   //Needed for roulette selection
float Jprob[100][3]; //Needed for roulette selection
int num_prob;         //Needed for roulette selection
float Ci[100][9];     //Needed for roulette selection
int clusters[100][500]; //Needed for Clustering
int n;   // number of nodes
int e;   // number of edges
int d;   // number of sinks
int demand; //demand of the flow
```

```
int num_paths; //number of paths in the paths array
int max_paths; //indicate the destination node with maximum
number of paths
int num_population; //number of elements of the population
int num_population_initial  ; //number of elements of the
population
int new_num_population; //number of  elements  of  the
population
int new_num_population1; //number  of  elements  of  the
population
int new_num_population2; //number  of  elements  of  the
population
int new_num_population3; //number  of  elements  of  the
population
int size_comparison_set; //size  of  the  comparison set
int maxNondominated;// maximum size of external
nondominated set

int f1_max_value,f1_min_value;//Are  the  bounds  of  the
function f1
int f2_max_value,f2_min_value;//Are  the  bounds  of  the
function f2

int del_path[MAX_NODES];//path to be removed by a constraint
int color[MAX_NODES]; // needed for breadth-first search
int pred[MAX_NODES];   // array to store augmenting path
int dest[MAX_NODES];   // array to store the sink nodes
float strength[MAX_NODES][2];  // array to store the
strength
int capacity[MAX_NODES][MAX_NODES];  //array to store the
link capacity
int delay[MAX_NODES][MAX_NODES];   //array to store the
link delay
int visited[MAX_NODES][MAX_NODES];   //array to store
information about the link visited in the paths search
int num_paths_by_destination[MAX_NODES][3]; //array to
indicate the number of paths by each destination
                                  // 0 : destination node
                                         // 1 : number
of paths to this destination
                                         // 2 : number
of the row where is the first position for this destination
```

```
int paths[MAX_NODES][MAX_NODES]; //array to store all paths
to all destinations
                                // 0 : destination node
                                // 1 to (-1) : nodes of
this path
                                // with (-1) values
identify the end of this path
int population[MAX_POPULATION][MAX_NODES][MAX_NODES];
//array to store the population
int population_branch[MAX_POPULATION][MAX_NODES]; //array
to store the population
int new_population[MAX_POPULATION][MAX_NODES][MAX_NODES];
//array to store the population
int new_population1[MAX_POPULATION][MAX_NODES][MAX_NODES];
//array to store the population for binary tournament
selection
int new_population2[MAX_POPULATION][MAX_NODES][MAX_NODES];
//array to store the population
int new_population3[MAX_POPULATION][MAX_NODES][MAX_NODES];
//array to store the population
int temp_population[MAX_POPULATION][MAX_NODES][MAX_NODES];
float fitness[MAX_POPULATION][2];//array to store fitness
of the chromosomes of the population
                                // 0 : hop count
                                // 1 : delay
int count_generations;
clock_t time1, time2, time3;

//A Queue for Breadth-First Search----------------------
int head,tail;
int q[MAX_NODES+2];

void enqueue (int x) {
    q[tail] = x;
    tail++;
    color[x] = GRAY;
}

int dequeue () {
    int x = q[head];
    head++;
    color[x] = BLACK;
```

```
        return x;
}
//----------------------------------------------------------
//Breadth-First Search--------------------------------------
int bfs (int start, int target) {
        int u,v;

        for (u=0; u<n; u++) {
        color[u] = WHITE;
        }
        head = tail = 0;
        enqueue(start);
        pred[start] = -1;
        while (head!=tail) {
   u = dequeue();
        // Search all adjacent white nodes v. If the capacity
         // from u to v in the residual network is positive,
            // enqueue v.
   for (v=0; v<n; v++) {
        if (color[v]==WHITE && visited[u][v] == 1) {
        enqueue(v);
        pred[v] = u;
         }
   }
        }
        // If the color of the target node is black now,
        // it means that we reached it.
        return color[target]==BLACK;
}
 int ai=0,aj=0;
#define MAX_PATHS 500

int s=0,ind=0,kk=1,target,a=1,kn=0;//Needed for
Backtracking search paths
int path[MAX_PATHS][MAX_NODES],ant[50][2];//Needed for
Backtracking search paths

//----------------------------------------------------------
void clear_branch(){//Clear branchs
        int i,j;
```

```
        s=0;ind=0;kk=1;a=1;kn=0;

        for(i=0;i<MAX_PATHS;i++){
            for(j=0;j<MAX_NODES;j++){
                path[i][j]=0;
            }
        }

        for(i=0;i<50;i++){
            for(j=0;j<2;j++){
                ant[i][j]=0;
            }
        }

        for(i=0;i<MAX_POPULATION;i++){
            for(j=0;j<MAX_NODES;j++){
                population_branch[i][j]=0;
            }
        }


        num_population_initial   =0;
        num_paths=0;
}
//------------------------------------------------------------
//--------Search Paths for Initial Population-------------
void update_population(){// Update backtracking population

        int w=0;
        while(path[num_population_initial   -1][w]!=-1){
            population_branch[num_population_initial-
1][w]=path[num_population_initial-1][w];
            w++;
        }
        if(path[num_population_initial-1][w]==-1){
            population_branch[num_population_initial-
1][w]=path[num_population_initial-1][w];
        }
    }
}
//------------------------------------------------------------
```

```
void move (int ii)
{   int u;
 for(u=ii;u<kk;u++)
 {
        path[num_paths-1][u]=path[num_paths-1][u+1];
 }
     path[num_paths-1][u]=-1;
}
//-----------------------------------------------------
void del_visited(int base){
 for(int ii=kk;ii>0;ii--)
 {
    if(visited[path[num_paths-1][ii-1]][path[num_paths-
1][ii]]!=1)
    {
       move (ii-1);
    }
 }

 while(ant[a-1][1]!=base)
{  ant[a-1][0]=ant[a][0];
    ant[a-1][1]=ant[a][1];
    ant[a][0]=0;
     ant[a][1]=0;
     a--;
 }
}
//-----------------------------------------------------
bool visit(int node,int c){

int sw=0;
for(int ii=0;ii<kn;ii++)
{
   if(path[c][ii]==node)
   sw=1;
}

if(sw==0)
return true;
else
```

```
return false;
}
//------------------------------------------------------------
update_path(int ii){//Update backtracking paths
   int jj=0;
  while((path[num_paths-1][jj]!=ii)&&(jj<MAX_NODES))
  {path[num_paths][jj]=path[num_paths-1][jj];
   jj++; kk++;
   }
  path[num_paths][jj]=path[num_paths-1][jj];
}
//------------------------------------------------------------
int Count(int i,int num_paths){

int cont=0;
int np=0;

np=num_paths;

while(path[np][cont]!=i)
{

   cont++;

}
return cont+1;
}


//------------------------------------------------------------
Search_Paths(int i)  //Backtracking search paths
{
          if(i<n)
          {
                    for(int j=ind;j<=n;j++)
                    {
                     if(visited[i][j]==1)
                     {

                      if(kk==1)
                     kn=Count(i,num_paths-1);
                     else
```

```
kn=Count(i,num_paths);

int np=0;

 if(kk>1)
 np=num_paths;
 else
 np=num_paths-1;

 if(visit(j,np))
    {
       if((kk==1)&&(num_paths>0))
       update_path(i);

     path[num_paths][kk]=j;

       if(j!=target)
        {
           ant[a][0]=i;
           ant[a][1]=j;
           a++;
           kk++;
           ind=0;
           Search_Paths(j);
        }else{
            path[num_paths][kk+1]=-1;
           num_population_initial   ++;
            num_paths++;
            ant[a][0]=i;
            ant[a][1]=j;
            a++;
            i=ant[a-1][0];
            ind=ant[a-1][1]+1;
            j=ind;
            ant[a][0]=0;
            ant[a][1]=0;
            a--;
            del_visited(i);
            update_population();
            kk=1;
```

```
                    }//end else
                  }//end visit
                }//end population
              }//end for
        }
}
//------------------------------------------------------------
void remove_path_2(int i_del){

        for(int i=i_del;i<num_population-1;i++){
            for(int h=0;h<d;h++){
              int j=0;
              while(population[i+1][h][j]!=-1)
                {
              population[i][h][j]=population[i+1][h][j];
              j++;
                }
                for(int k=j;k<MAX_NODES;k++){population
[i][h][k]=-1;}
                }
            }
        num_population--;
        int n=population[0][0][0];
}
//-------End Search Paths for Initial Population---------
void repetition(int
p[MAX_POPULATION][MAX_NODES][MAX_NODES],int ii,int
n){//Delete repeated paths
int v=0;

        for(int i=ii+1;i<num_population;i++){
            int v=0,k=0;
            for(int h=0;h<d;h++){

              int j=0;
              while (population[i][h][j] != -1) {

                if(population[i][h][j]==p[ii][h][j])
                v++;
                else
```

```
                 v=0;

                 j++;k++;
                   }
              }
               if(v==k)
               {remove_path_2(i);
                  i--;
                  }
          }
  }
//-------------------------------------------------------
void feasibility(int j){//feasibility Analysis

int fit=0;
int k=0,h;

for(h=0;h<d;h++){
while (population[j][h][k+1] != -1) {

 fit = fit + capacity[population[j][h][k]]
[population[j][h][k+1]];   //capacity
 k++;
 }

   if(fit<demand){
   remove_path_2(j);h=d;
   }
}
}
//initial population--------------------------------------
void initial_population() {

   int i,j,i1,j1,j2,k;
   int num_paths_by_destination_tmp[MAX_NODES][2];

   for (i=0; i<MAX_POPULATION; i++) {
           for(k=0;k<d;k++){
    for (j=0; j<MAX_POPULATION; j++) {
        new_population[i][k][j] = -1;
```

```
            new_population2[i][k][j] = -1;
            new_population3[i][k][j] = -1;
        }
                }
        }

    for(int ii=0;ii<=num_population;ii++)
    repetition(population,ii,num_population);

for(int i=0;i<num_population;i++)
feasibility(i);

i=0;
    while(population[i][0][0]!=-1 && population[i][0][1]!=0)
    {i++;
    }
    num_population=i;
    num_population_initial   =i;
}
//end initial_population()-------------------------------
//SetFit()-----------------------------------------------
void SetFit(int p,int i,  int s){

    strength[i][p]=s;
}
//end SetFit()-------------------------------------------
//GetFit()-----------------------------------------------
float GetFit(int p,int i){

    return strength[i][p];
}
//end GetFit()------------------------------------------
//dominatesOrEqual---------------------------------------
bool dominatesOrEqual(int i,int j,int
cam1[MAX_POPULATION][MAX_NODES][MAX_NODES],int
cam2[MAX_POPULATION][MAX_NODES][MAX_NODES]){

  int k=0,l,state1,state2,h;
  int win,loss,tie;
```

```
          int fit[2][2];
          fit[0][0]=0;
          fit[0][1]=0;
          fit[1][0]=0;
          fit[1][1]=0;

          for(h=0;h<d;h++){
          while (cam1[i][h][k+1] != -1) {
          fit[0][0] = fit[0][0]+1;//hop count
  fit[0][1] = fit[0][1] +
delay[cam1[i][h][k]][cam1[i][h][k+1]];//delay
 k++;
 }

             }
 k=0;

          for(h=0;h<d;h++){
          while (cam2[j][h][k+1] != -1) {
          fit[1][0] = fit[1][0]+1;//hop count
  fit[1][1] = fit[1][1] +
delay[cam2[j][h][k]][cam2[j][h][k+1]];//delay
 k++;
 }

             }
     win   = 0;
     loss = 0;
     tie = 0;
     if (fit[0][0] < fit[1][0])
         win++;
     else if (fit[0][0] == fit[1][0])
             tie++;
          else
             loss++;

     if (fit[0][1] < fit[1][1])
         win++;
     else if (fit[0][1] == fit[1][1])
             tie++;
          else
             loss++;
```

```
                    if(win>loss)
                    return true;
                    else
                       if(win==loss)
                            return true;
                            else
                            return false;


}
//end dominatesOrEqual------------------------------------
//calc. Strengths and fitness for non-dominated population-
------------------
void   calcStrengths()
{

   int    i, j;
   int    domCount,pop_count;

   for (i = new_num_population1 - 1; i >= 0; i--) {

      domCount = 0;
      pop_count=0;
      for (j = num_population - 1; j >= 0; j--){

         if(population[j][0][0]!=-1){
         pop_count++;
         if (dominatesOrEqual(i,j,new_population1,
population)){
 domCount++;
         }
         }
      }
      if(domCount==0)pop_count=domCount=1;
      SetFit(0,i, 1/(double(domCount) / double(pop_count +
1))); //num_population
   }
}
//end calcStrengths()------------------------------------
//calc.Fitness for dominated population------------------
void   calcFitness()
```

```
{

    int     i, j;
    double  fitness1;

    for (i = num_population - 1; i >= 0; i--) {

        fitness1 = 0.0;
      if(population[i][0][0]!=-1){
      for (j = new_num_population1 - 1; j >= 0; j--) {

        if (dominatesOrEqual(j,i,new_population1,population))
 fitness1 += GetFit(0,j);
      }
        SetFit(1,i, 1.0 + fitness1);
      }
    }
}
//end calcFitness()--------------------------------------
//Distance calculate-----------------------------------
double  IndDistance(int i,int j,int
cam1[MAX_POPULATION][MAX_NODES][MAX_NODES])
{
double distance;
distance=0;

            int fit[2][2],k=0,h;
            fit[0][0]=0;
            fit[0][1]=0;
            fit[1][0]=0;
            fit[1][1]=0;

            for(h=0;h<d;h++){
            k=0;
            while (cam1[i][h][k+1] != -1) {
            fit[0][0] = fit[0][0]+1;//hop count
 fit[0][1] = fit[0][1] +
delay[cam1[i][h][k]][cam1[i][h][k+1]];//delay
 k++;
 }
```

```
        }


        for(h=0;h<d;h++){
        k=0;
        while (cam1[j][h][k+1] != -1) {
        fit[1][0] = fit[1][0]+1;//hop count
 fit[1][1] = fit[1][1] +
delay[cam1[j][h][k]][cam1[j][h][k+1]];//delay
 k++;
 }
            }
   distance = sqrt(
                        (pow(fit[0][0] - fit[1][0], 2)) +
                        (pow(fit[0][1] - fit[1][1], 2))
                        );
   return distance;
}
//End distance-----------------------------------------------
-------------------
//Cluster distance------------------------------------------
-------------------
double  ClusterDistance(int  ic1, int  ic2)
{
   double  sum = 0;
   int  numOfPairs = 0;


   int  ik1=0;
   while(clusters[ic1][ik1]!=-1)//clusters[ic1][ik1]
   {
            int  ik2=0;
            while(clusters[ic2][ik2]!=-1){
             numOfPairs++;
             sum += IndDistance(ic1,ic2,new_population);
              ik2++;
            }
            ik1++;
   }
```

```
if(sum<0) sum=sum*-1;

return (sum / double(numOfPairs));

}
//End clusters distance--------------------------------
//Join Clusters----------------------------------------
void  JoinClusters(int  cluster1, int  cluster2, int&
numOfClusters)
{

int k=0;
while(clusters[cluster1][k]!=-1)
{
  k++;
}

int kk=0;
while(clusters[cluster2][kk]!=-1)
{clusters[cluster1][k]=clusters[cluster2][kk];
k++;kk++;
}

k=0;
while(clusters[cluster2][k]!=-1)
{clusters[cluster2][k]=-1;k++;}

numOfClusters--;
}
//End Join Clusters------------------------------------
//Find Cluster Centroid--------------------------------
int  ClusterCentroid(int cluster)
{

        int cluster_value[200][2],k=0,prom1=0,prom2=0;

        for(int i=0;i<100;i++)
        {cluster_value[i][0]=0;cluster_value[i][1]=0;}

        int num_cam=0;
```

```
        while(clusters[cluster][k]!=-1){
                int cl=clusters[cluster][k];
                int kk=0;
                for(int h=0;h<d;h++){
                while (new_population[cl][h][kk+1] != -1) {
                        cluster_value[cluster][0] =
cluster_value[cluster][0]+1;//hop count
                cluster_value[cluster][1]  = cluster_value
[cluster][1]+delay[new_population[cl][h][kk]][new_
population[cl][h][kk+1]];//delay
                kk++;
                }


                }


            prom1=cluster_value[cluster][0];
            prom2=cluster_value[cluster][1];
            num_cam++;
            k++;


        }
        prom1=prom1/num_cam;
        prom2=prom2/num_cam;

    for(int i=0;i<100;i++)
    {cluster_value[i][0]=0;cluster_value[i][1]=0;}
    int min_dist=1000,selected=0;
    float dist=0;
    k=0;

    while(clusters[cluster][k]!=-1){
    kk=0;
    int cl=clusters[cluster][k];
      cluster_value[cluster][0]=0;
      cluster_value[cluster][1]=0;

      for(int h=0;h<d;h++){
      while (new_population[cl][h][kk+1] != -1) {
                cluster_value[cluster][0] =
cluster_value[cluster][0]+1;//hop count
```

```
            cluster_value[cluster][1] = cluster_value
[cluster][1] +
delay[new_population[cl][h][kk]][new_population[cl][h][kk
+1]];//delay
            kk++;
    }
    }

     dist=sqrt(pow(cluster_value[cluster][0]-
prom1,2)+pow(cluster_value[cluster][1]-prom2,2));

     if(dist<min_dist)
     {
        selected=clusters[cluster][k];
        min_dist=dist;
     }
    k++;

  }

  clusters[cluster][0]=selected;
  k=1;
  while(clusters[cluster][k]!=-1){
        clusters[cluster][k]=-1;
        k++;
  }

  for(int h=0;h<d;h++){
  k=0;
  while(new_population[clusters[cluster][0]][h][k]!=-1){

new_population1[new_num_population1][h][k]=new_population
[clusters[cluster][0]][h][k];
  k++;
  }
  new_population1[new_num_population1][h][k]=-1;
  }
  new_num_population1++;
}
//End Cluster Centroid-------------------------------------
//Reduce Non-Dominated-------------------------------------
```

```
void   clustering(void)
{

int i;
for (i = 0 ; i < new_num_population; i++){
clusters[i][0]=i;
}

    int population_clusters=new_num_population;
    while(population_clusters>maxNondominated){

        float minDist=100;
        int join1,join2;
        int   numOfClusters = new_num_population;

        for (int   c = 0; c < numOfClusters; c++) {
        for (int   d = c + 1; d < numOfClusters; d++) {

         if(clusters[c][0]!=-1 && clusters[d][0]!=-1)
         {
                double dist=ClusterDistance(c,d);//
clusters[c][0],clusters[d][0]);

        if (dist < minDist) {
          minDist = dist;
           join1 = c;
           join2 = d;
                  }}
        }
        }

    JoinClusters(join1, join2, population_clusters);

    }
     int c=0,ind=0;
     while(c<population_clusters)
     {

     if(clusters[ind][0]!=-1)
     {
```

```
      ClusterCentroid(ind);
      c++;
      }
      ind++;
      }
}
//End Reduce Non-Dominated------------------------------
//Bubble_Sort-----------------------------------------
void Bubble_Sort()
{
   int   i,j;
   float temp1=0,temp2=0;

      for(i=0;i<new_num_population1;i++){
         for(j=i+1;j<new_num_population1;j++){

                  if(Prob[i][0]>Prob[j][0])
                  {
                     temp1=Prob[i][0];
                     temp2=Prob[i][1];
                     Prob[i][0]=Prob[j][0];
                     Prob[i][1]=Prob[j][1];
                     Prob[j][0]=temp1;
                     Prob[j][1]=temp2;


                  }
            }
         }

      for(i=0;i<num_prob;i++){
         for(j=i+1;j<num_prob;j++){

                  if(Prob[i][2]>Prob[j][2])
                  {
                     temp1=Prob[i][2];
                     temp2=Prob[i][3];
                     Prob[i][2]=Prob[j][2];
                     Prob[i][3]=Prob[j][3];
                     Prob[j][2]=temp1;
                     Prob[j][3]=temp2;
```

```
                    }
                }
            }
    }


//End Bubble_Sort-------------------------------------------
//Join_Prob-------------------------------------------------
void Join_Prob()
{
    int    temp1=0,temp2=0,i,j,ii=0;

    Jprob[0][0]=0;
    Ci[0][7]=0;
    Ci[0][8]=0;


        for(i=0;i<num_prob;i++){

            if(i==0){
            Jprob[i][0]=Prob[i][2];
            }else{
            Jprob[i][0]=Jprob[i-1][0]+Prob[i][2];
            }
            Jprob[i][1]=1;
            Jprob[i][2]=Prob[i][3];
        }
        j=i;
        for(i=0;i<new_num_population1;i++){
        Jprob[j][0]=Jprob[j-1][0]+Prob[i][0];
        Jprob[j][1]=0;
        Jprob[j][2]=Prob[i][1];
        j++;
        }

    }


//End Join_Prob---------------------------------------------
//Set_new_population----------------------------------------
void Set_new_population(int p,int  i)
```

```
{
int k,h;

if(p==0){

   for(h=0;h<d;h++){
   k=0;
   while(new_population1[i][h][k]!=-1)
   {

new_population[new_num_population][h][k]=new_population1[
i][h][k];
    k++;
   }
   new_population[new_num_population][h][k]=-1;
   }
   new_num_population++;
}else{

for(h=0;h<d;h++){
k=0;
   while(population[i][h][k]!=-1)
   {
    new_population[new_num_population][h][k]=
population[i][h][k];
    k++;
   }
   new_population[new_num_population][h][k]=-1;
   }
   new_num_population++;
}

}
//Set_new_population------------------------------------
//Roulette_Selection------------------------------------
void   Roulette_Selection()
{

     float num_random,sum_fitness=0,sum_ci=0,prob[100][3];
     int i,j,k;
```

```
for(i=0;i<MAX_POPULATION;i++)
{
for(k=0;k<MAX_NODES;k++){
    for(j=0;j<MAX_NODES;j++)
      {
        new_population[i][k][j]=-1;
      }
}
}

  new_num_population=0;

for(i=0;i<new_num_population1;i++){
sum_fitness=sum_fitness+GetFit(0,i);
Ci[i][0]=0;
}

for(i=0;i<num_population;i++){
if(population[i][0][0]!=-1){
sum_fitness=sum_fitness+GetFit(1,i);
Ci[i][1]=0;
}
}

for(i=0;i<new_num_population1;i++){
Ci[i][0]=1-double(GetFit(0,i)/sum_fitness);
sum_ci=sum_ci+Ci[i][0];
}

for(i=0;i<num_population;i++){
if(population[i][0][0]!=-1){
Ci[i][1]=1-double(GetFit(1,i)/sum_fitness);
sum_ci=sum_ci+Ci[i][1];
}
}

for(i=0;i<new_num_population1;i++){
Prob[i][0]=Ci[i][0]/sum_ci;
Prob[i][1]=i;
```

```
}

j=0;

for(i=0;i<num_population;i++){
if(population[i][0][0]!=-1){
Prob[j][2]=Ci[i][1]/sum_ci;
Prob[j][3]=i;
j++;

}
}

num_prob=j;

Bubble_Sort();
Join_Prob();

new_num_population=0;
for(i=0;i<(num_prob+new_num_population1)/2;i++){
num_random=rand()/32300.0;
int sw=0;
j=0;
while((j<num_population+new_num_population1-
1)&&(sw==0)){
    if(Jprob[0][0]>num_random){
    Set_new_population(Jprob[0][1],Jprob[0][2]);
    sw=1;
    }else{

if((Jprob[j+1][0]>=num_random)&&(Jprob[j][0]<num_random))
{
        Set_new_population(Jprob[j+1][1],Jprob[j+1][2]);
        sw=1;
    }
    j++;
    }
}
}
```

```
}
//end Roulette_Selection--------------------------------
void select_non_dominated()   {

        for (int i=0; i<MAX_POPULATION; i++) {
        for(int h=0;h<MAX_NODES;h++){
   for (int j=0; j<MAX_NODES; j++) {
      new_population[i][h][j] = -1;
   }
            }
          }

  int i,j,sw,k,l,h,state1,state2;
  int win,loss,tie;
  new_num_population = 0;
  for(i=0; i<num_population; i++) {
                j=0;
                sw=0;
                if(population[i][0][0]!=-1)
                {

                while((j<num_population)&&(sw==0)){
                win    = 0;
     loss = 0;
     tie = 0;

                if (fitness[i][0] < fitness[j][0])
        win++;
     else if (fitness[i][0] == fitness[j][0])
            tie++;
          else
             loss++;

     if (fitness[i][1] < fitness[j][1])
         win++;
     else if (fitness[i][1] == fitness[j][1])
            tie++;
          else
             loss++;
```

```
                      if (win > 0) {
        if (loss == 0) {
            state1 = NONDOMINATED;
            state2 = DOMINATED;
        }
        else {
            state1 = BOTH_NONDOMINATED;
            state2 = BOTH_NONDOMINATED;
        }
    }
    else {
        if (loss > 0) {
                                    sw=1;
            state1 = DOMINATED;
            state2 = NONDOMINATED;
        }
        else {
            state1 = BOTH_NONDOMINATED;
            state2 = BOTH_NONDOMINATED;
        }
    }

    if (tie == 4) {
        state1 = NONDOMINATED;
        state2 = DOMINATED;
    }
                    j++;
                }

                }

                        if(sw==0){
                                h=0;
                                while(h<d){
                                k=0;
                        while (population[i][h][k]
!= -1) {
                new_population[new_num_population][h][k]
= population[i][h][k];
```

```
                                        population[i][h][k]=
-1;
                k++;
            }
            new_population[new_num_population][h][k]  =
population[i][h][k];


                                population[i][h][k]=-1;
                k++;
                                        h++;
                                    }
        new_num_population++;

                            /* int jjj=i;
                                while(jjj<num_population-1){
                                for(h=0;h<d;h++){
                                    k=0;
                            while(population[jjj+1][h][k]!=-1){
                                        population[jjj][h][k]=
population[jjj+1][h][k];

                                    k++;
                                    }
                                population[jjj][h][k]=-1;
                            }

                            jjj++;
                            }
                            num_population--;
                                */
                            }

    }


        for(i=0; i<num_population; i++){
                if(population[i][0][0]==-1)
                {
                j=i;
                        while(j<num_population){
                        for(h=0;h<d;h++){
```

```
                                            k=0;
                            while(population[j+1][h][k]!=-1){
                                        population[j][h][k]=
population[j+1][h][k];
                                        k++;
                                        }
                                    population[j][h][k]=-1;
                            }
                            j++;
                            }
                            num_population--;
                    }
            }


  if (num_population == 1) {
      k=0;
      for (l=0; l < d; l++) {
          while (population[0][l][k] != -1) {
              new_population[0][l][k] =
population[0][l][k];
                                    population[0][l][k]=-1;
              k++;
          }
          new_population[0][l][k] = population[0][l][k];
                                    population[0][l][k]=-1;
          k++;
      }
      new_num_population++;
  }
}


//SPEA--------------------------------------------------
void SPEA()
{
int i,k,h;
    select_non_dominated();
  if (new_num_population > maxNondominated){
  new_num_population1=0;
  clustering();
```

```
  }else{
  new_num_population1=0;
  for(i=0;i<new_num_population;i++){
          for(h=0;h<d;h++){
          k=0;
          while(new_population[i][h][k]!=-1){

new_population1[new_num_population1][h][k]=new_population
[i][h][k];

          k++;
          }
          new_population1[new_num_population1][h][k]=-1;
          }
          new_num_population1++;
  }
  }
  calcStrengths();
  calcFitness();
}
//End SPEA------------------------------------------------
//Functions Calculate -----------------------------------
void Functions_calculate(int
pop[MAX_POPULATION][MAX_NODES][MAX_NODES],int num_pop) {

  int i,j,k,max,min,max2,min2;
          max=0;
          min=1000;
          max2=0;
          min2=1000;
  float alpha;

  for (i=0; i<num_pop; i++) {
      for (j=0; j<2; j++) {
          fitness[i][j] = 0.0;
      }

  }
  for (i=0; i<num_pop; i++) {
      k=0;
      for (j=0; j < d; j++) {
```

```
                        k=0;
               while (pop[i][j][k+1] != -1) {
                    fitness[i][0] = fitness[i][0] + 1;//hop
count
                    fitness[i][1] = fitness[i][1] +
delay[pop[i][j][k]][pop[i][j][k+1]];//delay
                    k++;
               }

               k++;
          }
     }


}
//end Functions calculate-------------------------------
----------------------
//end restriction------------------------------------------
-------------------
void restriction(int paths-tmp
[MAX_POPULATION][MAX_NODES][MAX_NODES],int num){
for(int i=0;i<num;i++){
for(int h=0;h<d;h++){
          int j=0,sw=0;
          while((paths-tmp[i][h][j]!=-1)&&(sw==0)){


                    if(capacity[paths-tmp[i][h][j]][paths-
tmp[i][h][j+1]]<demand)
                    {
                         if(capacity[paths-tmp[i][h][j]][paths-
tmp[i][h][j+1]]!=0){


                         for(int ii=i;ii<=num;ii++){

                         for(int hh=0;hh<d;hh++){
                           int jj=0;

                             while(paths-tmp[ii][hh][jj]!=-1)
                             {
```

```
                              paths-tmp[ii][hh][jj]=paths-
tmp[ii+1][hh][jj];
                        jj++;

                      }
                    }

                }

              sw=1;
              }
            }
            j++;  }
}
    }


}
//end restriction----------------------------------------
//cross-over pointer-------------------------------------
int crossover_position(int i,int des){
int sw=0,j=0;
while(sw==0){
        if((new_population[i][des][j]==
-1)||(new_population[i+1][des][j]==-1)){
            sw=1;
        }
        j++;
}
return j-2;
}
//end cross-over pointer---------------------------------
//verify_path-------------------------------------------
bool verify_path(int
pop[MAX_POPULATION][MAX_NODES][MAX_NODES], int indice){
    int sw;
    for(int h=0;h<d;h++){
    int j=0;sw=0;

            while(pop[indice][h][j+1]!=-1)
```

```
                {
                    if(capacity
[pop[indice][h][j]][pop[indice][h][j+1]]==0)
                    {sw=1;h=d;}
                j++;
                }
    }
    if(sw==0)
    return false;
    else
    return true;

}
//end verify_path----------------------------------------
//remove_path_3-----------------------------------------
void remove_path_3(int indice, int
pop[MAX_POPULATION][MAX_NODES][MAX_NODES],int size){

        int i=0,j,h;
        for(h=0;h<d;h++){
        for(i=indice;i<MAX_NODES;i++)
        {pop[indice][h][i]=-1;}
        }

        for(i=indice;i<size-1;i++){
        for(h=0;h<d;h++){
            j=0;
            while(pop[i+1][h][j]!=-1)
              {
            pop[i][h][j]=pop[i+1][h][j];
            j++;
            }
          for(int k=j;k<MAX_NODES;k++){pop[i][h][k]=-1;}
          }

        }

}
//end remove path----------------------------------------
//Repeated_nodes -----------------------------------------
```

```
bool Repeated_nodes(int
pop[MAX_POPULATION][MAX_NODES][MAX_NODES], int indice){
int i=0,j=0,sw=0,h;

for(h=0;h<d;h++){
while(pop[indice][h][i+1]!=-1)
{
    j=0;
    while(pop[indice][h][j]!=-1){

            if(i!=j){

                    if(pop[indice][h][i]==pop[indice][h][j]){
                        sw=1;
                        h=d;
                    }
            }
            j++;
            }
        i++;
}
}


if(sw==0)
return false;
else
return true;


}
//End Repeated_nodes -----------------------------------
//cross over -------------------------------------------
// Figure 4.33
void cross_over() {

for (int i=0; i<MAX_POPULATION; i++) {
        for(int h=0;h<d;h++){
    for (int j=0; j<MAX_POPULATION; j++) {
        new_population2[i][h][j] = -1;
```

```
    }
        }
    }

int i,l,k,k1,k2,k3,cross_over_position,dest_rand;
float num_random;
new_num_population2 = 0;

        if(new_num_population == 1){

                for(int h=0;h<d;h++){
                k3=0;
                while(new_population[0][h][k3] != -1){

new_population2[0][h][k3]=new_population[0][h][k3];
                k3++;
                }

new_population2[0][h][k3]=new_population[0][h][k3];
                }
        }else{

        for(i=0; i<new_num_population-1; i++){
        num_random = rand()/32300.0;

        if (num_random <= CROSS_OVER_PROBABILITY) {
        for(int h=0;h<d;h++){

                cross_over_position=
random(crossover_position(i,h));

                if(cross_over_position==0){
                cross_over_position=1;
                }

                k=0;k1=0;k2=0;
                while(k<cross_over_position){
            new_population2[new_num_population2][h][k]
= new_population[i][h][k];
            new_population2[new_num_population2+1][h][k]
= new_population[i+1][h][k];
```

```
                    k++;
                    }
                    k1=k;
                    k2=k;
                    while(new_population[i+1][h][k1]!= -1){

new_population2[new_num_population2][h][k1]=new_population
[i+1][h][k1];
                    k1++;
                    }

new_population2[new_num_population2][h][k1]=new_population
[i+1][h][k1];

                    while(new_population[i][h][k2]!= -1){

new_population2[new_num_population2+1][h][k2]=new_population
[i][h][k2];
                    k2++;
                    }

new_population2[new_num_population2+1][h][k2]=new_populat
ion[i][h][k2];

              }
              new_num_population2=new_num_population2+2;
              }
              }


              }


              for(i=0; i<new_num_population; i++){
                  for(int h=0;h<d;h++){
                  int k4=0;
                  while(new_population[i][h][k4]!= -1){

new_population2[new_num_population2][h][k4]=new_population
[i][h][k4];
                  k4++;
                  }
```

```
new_population2[new_num_population2][h][k4]=new_population
[i][h][k4];
                        k4++;
                        }
                    new_num_population2++;
              }


      for(k=0;k<new_num_population2;k++)
     {
    if(Repeated_nodes(new_population2,k))      // repeated
remove
    {remove_path_3(k,new_population2,new_num_population2);
       new_num_population2--;
       k--;
      }else{
          if(verify_path(new_population2,k))     // Path
Feasible

{remove_path_3(k,new_population2,new_num_population2);
              new_num_population2--;
              k--;
             }
        }

     }


}
//end cross_over()----------------------------------------
//mutation position--------------------------------------
int m_position(int i,int des){
int sw=0,j=0;
while(sw==0){
          if(new_population[i][des][j]==-1){
                sw=1;
          }
          j++;
}
return j-2;
}
```

```
//end_mutation_position--------------------------------
//mutation---------------------------------------------
// Figure 4.35
void mutation() {

for (int i=0; i<MAX_POPULATION; i++) {
        for(int h=0;h<d;h++){
    for (int j=0; j<MAX_POPULATION; j++) {
        new_population3[i][h][j] = -1;
    }
        }
    }
        int
i,l,k,k1,k2,u,i2,mutation_position,p,mutation_rand,h;
 float num_random;
        int paths_tmp[MAX_NODES];

 new_num_population3 = 0;

        for(i=0; i<new_num_population2-1; i++){

        num_random = rand()/32300.0;

            if (num_random <= MUTATION_PROBABILITY){

                mutation_rand=random(d);

                mutation_position=
random(m_position(i,mutation_rand));

                if(mutation_position<=1){
                mutation_position=2;
                }
                k=0;k1=0;

                    while(k<mutation_position){

new_population3[new_num_population3][mutation_rand][k] =
new_population2[i][mutation_rand][k];
                        k++;
                        }
```

```
if(bfs(new_population2[i][mutation_rand][k],dest[mutation
_rand])){

                        paths_tmp[0] = dest[mutation_rand];
          for (u=dest[mutation_rand]; pred[u]>=0;
u=pred[u]) {
                                    k1++;
              paths_tmp[k1] = pred[u];
                            visited[pred[u]][u] = 0;
          }




new_population3[new_num_population3][mutation_rand][k]  =
pred[u];

          for (int i1 = k1; i1>=0; i1--) {

new_population3[new_num_population3][mutation_rand][k]=
paths_tmp[i1];
          k++;
          }

new_population3[new_num_population3][mutation_rand][k]=
-1;

                        for(h=0;h<d;h++){

                          if(h!=mutation_rand)
                          {
                              k=0;
                        while(new_population2[i][h][k]!=
-1){

new_population3[new_num_population3][h][k]=new_population
2[i][h][k];
                                k++;
                                }

new_population3[new_num_population3][h][k]=-1;
                                }
                        }
```

```
                                new_num_population3++;
                                }
                    }
               }


        for(k=0;k<new_num_population3;k++)
        {
        if(Repeated_nodes(new_population3,k))
       {remove_path_3(k,new_population3,new_num_population3);
          new_num_population3--;
          k--;
          }else{
               if(verify_path(new_population3,k))

  {remove_path_3(k,new_population3,new_num_population3);
                  new_num_population3--;
                  k--;
                  }
           }
        }
 }
 //end mutation()------------------------------------------
 void Clear_cluster(){

    for(int i=0;i<100;i++){
    for(int j=0;j<500;j++){
    clusters[i][j]=-1;
    }
    }


 }
 //UpDate new populations-------------------------------
 void update_new_population() {

  int i,j,k,l;

            int num_temp_population=0;

            for (i=0; i<MAX_POPULATION; i++) {
```

```
            for(l=0;l<d;l++){
       for (j=0; j<MAX_POPULATION; j++) {
                   temp_population[i][l][j] = -1;
                   }
                   }
   }


            int ii=0;
   for (i=0; i<num_population; i++) {

                   if(population[i][0][0]!=-1){
                     for(l=0;l<d;l++){
                     j=0;
       while(population[i][l][j]!=-1){

                           temp_population[ii][l][j]=
   population[i][l][j];

                           j++;

                     }
                     temp_population[ii][l][j]=-1;
                     }
                     ii++;
                   }
   }

            num_temp_population=ii;

            for (i=0; i<MAX_POPULATION; i++) {
            for(l=0;l<d;l++){
            for (j=0; j<MAX_POPULATION; j++)
   {population[i][l][j]=-1;}}}

            num_population = 0;
            for(i=0;i<new_num_population1;i++){

              if(new_population1[i][0][j]!=-1)
              {
```

```
            for(l=0;l<d;l++){
            j=0;
        while(new_population1[i][l][j]!=-1){
                    popula-
tion[num_population][l][j]=new_population1[i][l][j];
                    j++;
                }
                population[num_population][l][j]=-1;
            }

                num_population++;
            }

            }

        for(i=0; i<new_num_population2; i++) {
                for(l=0;l<d;l++){
                k=0;
        while (new_population2[i][l][k] != -1) {

        population[num_population][l][k]  =
new_population2[i][l][k];
        k++;
        }
        population[num_population][l][k]  =
new_population2[i][l][k];
        k++;
                }
                num_population++;

    }

    for(i=0; i<new_num_population3; i++) {
                for(l=0;l<d;l++){
                k=0;
        while (new_population3[i][l][k] != -1) {
        population[num_population][l][k]  =
new_population3[i][l][k];
        k++;
        }
        population[num_population][l][k]  =
new_population3[i][l][k];
        k++;
```

```
                    }
                    num_population++;

    }



        for (i=0; i<MAX_POPULATION; i++) {
        for(l=0;l<d;l++){
    for (j=0; j<MAX_POPULATION; j++) {
                    new_population[i][l][j]  = -1;
                    new_population1[i][l][j] = -1;
                    new_population2[i][l][j] = -1;
                    new_population3[i][l][j] = -1;
                    temp_population[i][l][j] = -1;
                    }
                    }

    }


    Clear_cluster();
}
//end update_new_population()----------------------------
//population_view()-------------------------------------
void population_view(int
population_v[MAX_POPULATION][MAX_NODES][MAX_NODES], int
size_population,int count_generations ,String title) {
//char *title
 int i,k,l,g,h;

 printf("\n\n%s\n\n",title);
         printf("GENERATION  \tPOPULATION\n");
 for(i=0; i < size_population; i++) {
     k=0;
                    printf("%d \t\t",count_generations);
                        for(l=0;l<d;l++){
                        k=0;
         while (population_v[i][l][k] != -1) {
             printf("%d,",population_v[i][l][k]);
                                k++;

        }
```

```
                                        printf("\n\t\t");
                                        }
                                        printf("\n");
            k++;
  }
}
//end population_view()----------------------------------
//FITNESS VIEW----------------------------------------
void fitness_view(int tam,int generation) {
 int i;

 printf("\FUNCTION VALUES\n\n");
        printf("CHROMOSOME \t FUNCTIONS \t GENERATION\n");
 for(i=0; i < tam; i++) {

     printf("Chromosome %d \t Hop Count \t %f\n",i,
fitness[i][0]);
     printf("Chromosome %d \t Delay      \t %f\n",i,
fitness[i][1]);
         }
 printf("\n");


}
//end fitness_view()------------------------------------
//--------------------------------------------------------
//--------------------------------------------------------
void B_INPUTFILE()
{
int a,b,c,h,i,j,l;
     FILE* input = fopen("mf_MOEA.in","r");
     // read number of nodes, edges and sinks
     fscanf(input,"%d %d %d\n",&n,&e,&d);

     //printf("%d %d\n",n,e);
     // initialize empty capacity matrix
     for (i=0; i<n; i++) {
     num_paths_by_destination[i][0] = 0;
     num_paths_by_destination[i][1] = 0;
  for (j=0; j<n; j++) {
      capacity[i][j] = 0;
```

```
    delay[i][j] = 0;
    visited[i][j] = 0;
}
    }
    for (i=0; i<MAX_POPULATION; i++) {
    for(l=0;l<MAX_NODES;l++){
for (j=0; j<MAX_NODES; j++) {
     population[i][l][j] = 0;
}
        }
    }

    // read sink nodes
printf("SINKS NODES:\t");
for (i=0; i<d; i++) {
    fscanf(input,"%d",&dest[i]);
        printf("%d , ",dest[i]);

                  target=dest[i];
}
// read demand of the multicast flow (kbps)
printf("\nDEMAND FLOW:\t");
fscanf(input,"%d\n",&demand);

      printf("%d Kbps\n\n",demand);
    // read edge capacities
printf("EDGE CAPACITIES\n");

    for (i=0; i<e; i++) {
    fscanf(input,"%d %d %d %d",&a,&b,&c,&h);
     printf("%d %d %d %d\n",a,b,c,h);

            capacity[a][b] = c;
    delay[a][b] = h;
    visited[a][b] = 1;
    }

    int r=0;
    r=visited[8][12];
```

```
      fclose(input);
}
//-----------------------------------------------------------
void OUTFILE(int size_population,int
population_v[MAX_POPULATION][MAX_NODES][MAX_NODES]){
FILE* output = fopen("OUTPUT_FILE.txt","w");
      // read number of nodes, edges and sinks
      fprintf(output,"%s\n\n","FINAL POPULATION");
      fprintf(output,"GENERATION \tPOPULATION\n\n");
         int i,k,l;
  for(i=0; i < size_population; i++) {


                                    fprintf(output,"%d \t\
t",count_generations);
                                    for(l=0;l<d;l++){
                                    k=0;
         while (population_v[i][l][k] != -1) {
             fprintf(output,"%d,",population_v[i][l][k]);
                                       k++;
         }
                                    fprintf(output,"\n\t\t");
                                    }
                                    fprintf(output,"\n\n");
         k++;
  }


         fprintf(output,"\nFUNCTION VALUES\n\n");
         fprintf(output,"CHROMOSOME \t FUNCTIONS \t
GENERATION\n\n");
  for(i=0; i < size_population; i++) {
      fprintf(output,"Chromosome %d \t Hop Count\t %f\
n",i,fitness[i][0]);
      fprintf(output,"Chromosome %d \t Delay    \t %f\
n",i,fitness[i][1]);


      }


  fprintf(output,"\n");
         fclose(output);
}
//-----------------------------------------------------------
```

```
void sort_best(){
int i,j,k,salt1,salt2,temp;
for(i=0;i<num_population_initial   ;i++){
        k=0;
        salt1=0;
        while(population_branch[i][k]!=-1){
        salt1++;
        k++;
        }
for(j=i+1;j<num_population_initial;j++){
        k=0;
        salt2=0;
        while(population_branch[j][k]!=-1){
        salt2++;
        k++;
        }

        if(salt2<salt1){
        k=0;
        while(k<=salt1){
        temp=population_branch[i][k];
        population_branch[i][k]=population_branch[j][k];
        population_branch[j][k]=temp;

        k++;
        }
        salt1=salt2;
        population_branch[j][k]=-1;
        population_branch[i][k]=-1;
        }
}
}
}
//-----------------------------------------------------------
void Search_trees(){
   int h,k,i,npi=1000;

       for(h=0;h<d;h++){
           target=dest[h];
           Search_Paths(0);    // Source node : node 0
```

```
        sort_best();

        for(i=0;i<num_population_initial   ;i++){
                k=0;
                while(population_branch[i][k]!=-1)
                {
            population[i][h][k]=population_branch[i][k];
                k++;
                }
                population[i][h][k]=-1;
        }

        if(npi>num_population_initial)
        npi=num_population_initial;

      num_population_initial=0;
      clear_branch();
    }

    num_population_initial=npi;
}
//--------------------------------------------------------

int main ()
{
        B_INPUTFILE();
        maxNondominated = MAX_NONDOMINATED_SOLUTIONS;
        int  nondominatedSetUpperBound = num_population
+ maxNondominated;

        time1 = clock();
        Search_trees();
  initial_population();

        population_view(population,
num_population_initial,0,"INITIAL POPULATION");
        num_population=num_population_initial-1;
  count_generations = 0;

  while (count_generations < MAX_GENERATIONS) {
```

```
     //printf("\n\nGeneration: %d\n\
n",count_generations+1);
     Functions_calculate(population,num_population);
     //fitness_view(tringGrid10,count_generations);
                    Clear_cluster();
                    SPEA();
                    Roulette_Selection();
     cross_over();

population_view(new_population2,new_num_population2,count
_generations,"POPULATION AFTER CROSSOVER");
     mutation();

population_view(new_population3,new_num_population3,count
_generations,"POPULATION AFTER MUTATION   ");
     update_new_population();
                    population_view(popula-
tion,num_population,count_generations,"NEW POPULATION");
     count_generations++;
  }


        for(int ii=0;ii<num_population;ii++)
        repetition(population,ii,num_population);

        Functions_calculate(population,num_population);
        select_non_dominated();

population_view(new_population,new_num_population,4,
"FINAL POPULATION");

Functions_calculate(new_population,new_num_population);
      fitness_view(new_num_population,count_generations);
          time2 = clock();
 printf("\nMOEA running time : %f \n",(float)(time2-
time1)/CLK_TCK*1000);
          OUTFILE(new_num_population,new_population);
 getch();
     return 0;
}
//-------------------------------------------------------
```

Input File (mf_MOEA.in)

```
14 40 2  /* Nodes      Links     Destination Nodes */
8                /* Destination node 1 */
12               /* Destination node 2 */
256              /* Demand of the flow */
0 1 1536 9/* Node i    Node j   Link Capacity    Delay */
0 2 1536 9
0 3 1536 7
1 0 1536 9
2 0 1536 9
3 0 1536 7
1 3 1536 13
1 6 1536 20
3 1 1536 13
6 1 1536 20
2 4 1536 7
2 7 1536 26
4 2 1536 7
7 2 1536 26
3 10 1536 15
10 3 1536 15
4 5 1536 7
4 10 1536 11
5 4 1536 7
10 4 1536 11
5 6 1536 7
6 5 1536 7
6 9 1536 7
9 6 1536 7
7 8 1536 5
8 7 1536 5
7 13 1536 8
13 7 1536 8
8 9 1536 5
9 8 1536 5
8 12 1536 7
12 8 1536 7
9 11 1536 8
11 9 1536 8
10 12 1536 14
12 10 1536 14
10 11 1536 9
```

```
11  10  1536  9
12  13  1536  4
13  12  1536  4
```

Example 1

Destination nodes: 8 and 12

SOURCE	0
TARGET	8,12

SOL	CHROMOSOME	FUNCTIONS	
		HOP	DELAY
	0,2,7,8		
1	0,3,10,12	6	76

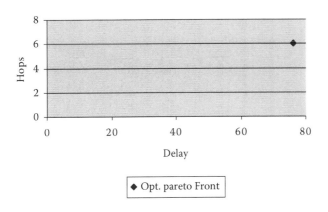

Delay

◆ Opt. pareto Front

Topology

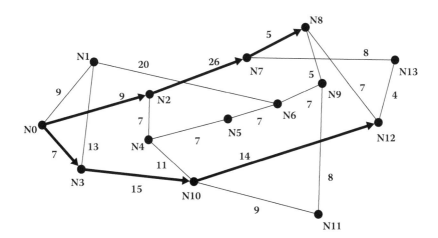

Example 2

Destination nodes: 8 and 13

SOURCE	0
TARGET	8,13

SOL	CHROMOSOME	FUNCTIONS	
		HOP	DELAY
	0,2,7,8		
1	0,2,7,13	6	83
	0,1,6,9,8		
2	0,3,10,12,13	8	81

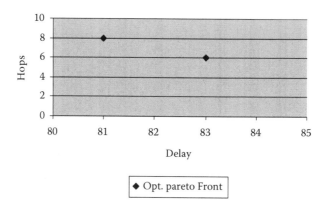

♦ Opt. pareto Front

First solution:

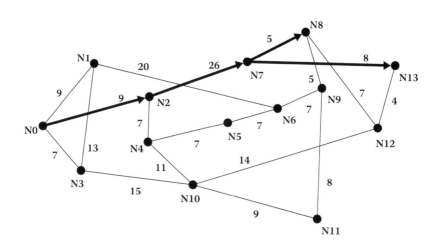

Second solution:

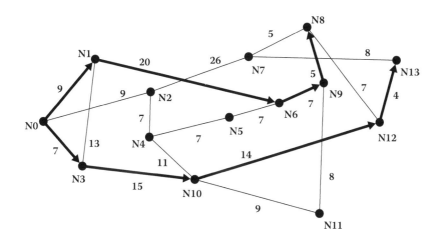

Program in C to Section 4.7.1 (Adding Several Flows *f*)

```
//------------------------------------------------------
/*

Program: MOEA - UNICAST - MULTI FLOWS
*/

#include <vcl.h>
#pragma hdrstop

#include <stdlib.h>
#include <math.h>
#include <stdio.h>
#include <conio.h>
#include <time.h>
//------------------------------------------------------
//Basic Definitions
#define WHITE 0
#define GRAY 1
#define BLACK 2
#define MAX_NODES 50
#define MAX_FLOWS 3
#define MAX_POPULATION 400
```

```
#define MAX_GENERATIONS 5
#define SIZE_COMPARISON 1.0
#define CROSS_OVER_PROBABILITY 0.85
#define MUTATION_PROBABILITY 0.2
#define ERROR 0.00001
#define oo 1000000000
#define oof 10000.0
#define zero 0.000001
#define DOMINATED 0
#define NONDOMINATED 1
#define BOTH_NONDOMINATED 2
#define MAX_NONDOMINATED_SOLUTIONS 3
//

//Declarations
float Prob[MAX_POPULATION][4];   //Needed for roulette
selection
float Jprob[MAX_POPULATION][3]; //Needed for roulette
selection
int num_prob;           //Needed for roulette selection
float Ci[MAX_POPULATION][9];      //Needed for roulette
selection
int clusters[MAX_POPULATION][MAX_POPULATION]; //Needed for
Clustering
int n;   // number of nodes
int e;   // number of edges
int d;   // number of sinks
int flows;   // number of flows
int demand[MAX_FLOWS]; //demand of the flow
int num_paths; //number of paths in the paths array
int max_paths; //indicate the destination node with maximum
number of paths
int num_population; //number of elements of the population
int num_population_initial  ; //number of elements of the
population
int new_num_population; //number of elements of the
population
int new_num_population1; //number of elements of the
population
int new_num_population2; //number of elements of the
population
```

```
int new_num_population3; //number of elements of the
population
int size_comparison_set; //size of the comparison set
int maxNondominated;// maximum size of external
nondominated set


int f1_max_value,f1_min_value;//Are the bounds of the
function f1
int f2_max_value,f2_min_value;//Are the bounds of the
function f2


int del_path[MAX_NODES];//path to be removed by a constraint
int color[MAX_NODES]; // needed for breadth-first search
int pred[MAX_NODES];   // array to store augmenting path
int dest[MAX_NODES];   // array to store the sink nodes
float strength[MAX_NODES][2];   // array to store the
strength
int capacity[MAX_NODES][MAX_NODES];   //array to store the
link capacity
int temp_capacity[MAX_NODES][MAX_NODES];  //array to store
the temporal link capacity
int delay[MAX_NODES][MAX_NODES];   //array to store the
link delay
int visited[MAX_NODES][MAX_NODES];   //array to store
information about the link visited in the paths search
int num_paths_by_destination[MAX_NODES][3]; //array to
indicate the number of paths by each destination
                                 // 0 : destination node
                                        // 1 : number
of paths to this destination
                                        // 2 : number
of the row where is the first position for this destination
int paths[MAX_NODES][MAX_NODES]; //array to store all paths
to all destinations
                                 // 0 : destination node
                                 // 1 to (-1) : nodes of
this path
  /*                             // with (-1) values
identify the end of this path
int population[MAX_POPULATION][MAX_NODES][MAX_NODES];
//array to store the population
int population_branch[MAX_POPULATION][MAX_NODES]; //array
to store the population
```

```
int new_population[MAX_POPULATION][MAX_NODES][MAX_NODES];
//array to store the population
int new_population1[MAX_POPULATION][MAX_NODES][MAX_NODES];
//array to store the population for binary tournament
selection
int new_population2[MAX_POPULATION][MAX_NODES][MAX_NODES];
//array to store the population
int new_population3[MAX_POPULATION][MAX_NODES][MAX_NODES];
//array to store the population
int temp_population[MAX_POPULATION][MAX_NODES][MAX_NODES];
*/

int population[MAX_POPULATION][MAX_FLOWS][MAX_NODES];
//array to store the population
int first_population[MAX_POPULATION][MAX_NODES];
int new_population[MAX_POPULATION][MAX_FLOWS][MAX_NODES];
//array to store the population
int new_population1[MAX_POPULATION][MAX_FLOWS][MAX_NODES];
//array to store the population for binary tournament
selection
int new_population2[MAX_POPULATION][MAX_FLOWS][MAX_NODES];
//array to store the population
int new_population3[MAX_POPULATION][MAX_FLOWS][MAX_NODES];
//array to store the population
int temp_population[MAX_POPULATION][MAX_FLOWS][MAX_NODES];


float fitness[MAX_POPULATION][3];//array to store fitness
of the chromosomes of the population
                                // 0 : hop count
                                // 1 : delay
int count_generations;
clock_t time1, time2, time3;

//A Queue for Breadth-First Search----------------------
int head,tail;
int q[MAX_NODES+2];

void enqueue (int x) {
    q[tail] = x;
    tail++;
    color[x] = GRAY;
```

```
}

int dequeue () {
    int x = q[head];
    head++;
    color[x] = BLACK;
    return x;
}
//----------------------------------------------------------
//Breadth-First Search-------------------------------------
int bfs (int start, int target) {
    int u,v;

    for (u=0; u<n; u++) {
    color[u] = WHITE;
    }
    head = tail = 0;
    enqueue(start);
    pred[start] = -1;
    while (head!=tail) {
 u = dequeue();
      // Search all adjacent white nodes v. If the capacity
       // from u to v in the residual network is positive,
          // enqueue v.
  for (v=0; v<n; v++) {
      if (color[v]==WHITE && visited[u][v] == 1) {
      enqueue(v);
      pred[v] = u;
       }
 }
    }
    // If the color of the target node is black now,
    // it means that we reached it.
    return color[target]==BLACK;
}
 int ai=0,aj=0;
#define MAX_PATHS 500

int s=0,ind=0,kk=1,target,a=1,kn=0;//Needed for
Backtracking search paths
```

```
int path[MAX_PATHS][MAX_NODES],ant[50][2];//Needed for
Backtracking search paths

//----------------------------------------------------------
void clear_branch(){//Clear branchs
 /*     int i,j;
        s=0;ind=0;kk=1;a=1;kn=0;

        for(i=0;i<MAX_PATHS;i++){
            for(j=0;j<MAX_NODES;j++){
                path[i][j]=0;
            }
        }

        for(i=0;i<50;i++){
            for(j=0;j<2;j++){
                ant[i][j]=0;
            }
        }

        for(i=0;i<MAX_POPULATION;i++){
            for(j=0;j<MAX_NODES;j++){
                population_branch[i][j]=0;
            }
        }


        num_population_initial   =0;
        num_paths=0;*/
}
//----------------------------------------------------------
//-------Search Paths for Initial Population------------
void update_population(){// Update backtracking population

        int w=0;
        while(path[num_population_initial-1][w]!=-1){
            first_population[num_population_initial-
1][w]=path[num_population_initial-1][w];
            w++;
        }
```

```
      if(path[num_population_initial-1][w]==-1){
           first_population[num_population_initial-
1][w]=path[num_population_initial-1][w];
    }
 }
//---------------------------------------------------------
void move   (int  ii)
{   int  u;
 for(u=ii;u<kk;u++)
 {
       path[num_paths-1][u]=path[num_paths-1][u+1];
 }
     path[num_paths-1][u]=-1;
}
//---------------------------------------------------------
void del_visited(int base){
 for(int  ii=kk;ii>0;ii--)
 {
    if(visited[path[num_paths-1][ii-1]][path[num_paths-
1][ii]]!=1)
    {
       move(ii-1);
     }
 }

 while(ant[a-1][1]!=base)
{  ant[a-1][0]=ant[a][0];
    ant[a-1][1]=ant[a][1];
    ant[a][0]=0;
     ant[a][1]=0;
     a--;
 }
}
//---------------------------------------------------------
bool visit(int node,int c){

int  sw=0;
for(int  ii=0;ii<kn;ii++)
{
   if(path[c][ii]==node)
```

```
    sw=1;
}

if(sw==0)
return true;
else
return false;
}

//------------------------------------------------------------
//Sort Paths by Hop Count
void initial_sort(){
int i,j,k,hops1,hops2,temp;
hops1=0;hops2=0;
for(i=0;i<num_population_initial   ;i++){
k=0;
while(first_population[i][k]!=-1){
k++;
}
hops1=k;
for(j=i+1;j<num_population_initial;j++){
k=0;
while(first_population[j][k]!=-1){
k++;
}
hops2=k;

if(hops1>hops2)
{ k=0;
  while(k<hops1)
  {
     temp=first_population[i][k];
     first_population[i][k]=first_population[j][k];
     first_population[j][k]=temp;
     k++;
  }
  first_population[i][k]=-1;
  first_population[j][k]=-1;
  hops1=hops2;
}
```

```
}
}
}
//
//----------------------------------------------------------
----------------------
update_path(int ii){//Update backtracking paths
   int jj=0;
 while((path[num_paths-1][jj]!=ii)&&(jj<MAX_NODES))
 {path[num_paths][jj]=path[num_paths-1][jj];
   jj++; kk++;
   }
 path[num_paths][jj]=path[num_paths-1][jj];
}
//----------------------------------------------------------
int Count(int i,int num_paths){

int cont=0;
int np=0;

np=num_paths;

while(path[np][cont]!=i)
{
   cont++;

}
return cont+1;
}


//----------------------------------------------------------
Search_Paths(int i)  //Backtracking search paths
{
        if(i<n)
        {
                for(int j=ind;j<=n;j++)
                {
                 if(visited[i][j]==1)
                 {
```

```
   if(kk==1)
kn=Count(i,num_paths-1);
else
kn=Count(i,num_paths);

int np=0;

 if(kk>1)
 np=num_paths;
 else
 np=num_paths-1;

 if(visit(j,np))
    {
       if((kk==1)&&(num_paths>0))
       update_path(i);

     path[num_paths][kk]=j;

       if(j!=target)
         {
            ant[a][0]=i;
            ant[a][1]=j;
            a++;
            kk++;
            ind=0;
            Search_Paths(j);
         }else{
             path[num_paths][kk+1]=-1;
             num_population_initial   ++;
             num_paths++;
             ant[a][0]=i;
             ant[a][1]=j;
             a++;
             i=ant[a-1][0];
             ind=ant[a-1][1]+1;
             j=ind;
             ant[a][0]=0;
             ant[a][1]=0;
             a--;
```

```
                                  del_visited(i);
                                  update_population();
                                  kk=1;
                               }//end else
                           }//end visit
                       }//end population
                  }//end for
          }
}
//-----------------------------------------------------
void remove_path_2(int i_del){

        for(int i=i_del;i<num_population-1;i++){
           for(int h=0;h<flows;h++){
             int j=0;
             while(population[i+1][h][j]!=-1)
              {
             population[i][h][j]=population[i+1][h][j];
             j++;
             }
              population[i][h][j]=-1;j++;
              for(int k=j;k<MAX_NODES;k++){popula-
tion[i][h][k]=-1;}
            }
          }
        num_population--;
        int n=population[0][0][0];
}
//-------End Search Paths for Initial Population---------
void repetition(int
p[MAX_POPULATION][MAX_FLOWS][MAX_NODES],int ii,int
n){//Delete repeated paths
int v=0;

        for(int i=ii+1;i<num_population;i++){
            int v=0,k=0;
           for(int h=0;h<flows;h++){

            int j=0;
            while (population[i][h][j] != -1) {
```

```
            if(population[i][h][j]==p[ii][h][j])
            v++;
            else
            v=0;

            j++;k++;
            }
         }
         if(v==k)
         {
         remove_path_2(i);
            i--;
            }
      }
  }
//----------------------------------------------------------
void feasibility(int j){//feasibility Analysis

int fit=0,sw=0;
int k=0,h;

for(h=0;h<flows;h++){
while ((population[j][h][k+1] != -1)&&(sw==0)) {

 fit = fit + capacity[population[j][h][k]]
[population[j][h][k+1]];   //capacity
 k++;

         if(fit<demand[h]){
         remove_path_2(j);h=flows;sw=1;
         }

 }


}
}
//verify capacity----------------------------------------
bool verify_capacity(int i,int j,int k){
   bool sw=true;
```

```
   if(j!=-1)
   {
   if(temp_capacity[i][j]>=demand[k])
   {sw=true;
   temp_capacity[i][j]=temp_capacity[i][j]-demand[k];
   }
   else
   sw=false;
   }


   return sw;
}
//temp_capacity-------------------------------------------
void restart_capacity(){
for (int i=0; i<n; i++) {
 for (int j=0; j<n; j++) {
     temp_capacity[i][j] = capacity[i][j];
 }
     }
}
//initial population-------------------------------------
void initial_population() {

   restart_capacity();



   int i,j,i1,j1,j2,k,sh;
   int num_paths_by_destination_tmp[MAX_NODES][2];



   for (i=0; i<MAX_POPULATION; i++) {
        for(k=0;k<flows;k++){
   for (j=0; j<MAX_POPULATION; j++) {
       new_population[i][k][j] = -1;
       new_population2[i][k][j] = -1;
       new_population3[i][k][j] = -1;
   }
           }
   }
```

```
j1=0;
for (i=0; i<num_population_initial  /4; i++) {
        for(k=0;k<flows;k++){
        j=0;
                while(first_population[i][j]!=-1){
            population[j1][k][j]=first_population[i][j];
                    j++ ;
                }
                population[j1][k][j]=-1;
        }
        j1++;
}




if(flows>1){
for (i=0; i<num_population_initial  /4; i++) {
j2=i;
        while(j2<num_population_initial/4){
                for(k=0;k<flows;k++){
                j=0;
                if(k==0){
                    while(first_population[i][j]!=-1){
                        popula-
tion[j1][k][j]=first_population[i][j];
                        j++ ;
                    }
                    population[j1][k][j]=-1;
                }else{
                    while(first_population[j2][j]!=-1){
                        population
[j1][k][j]=first_population[j2][j];
                        j++ ;
                    }
                    population[j1][k][j]=-1;
                }
```

```
                    j2++;
                    }
                    j1++;
            }
   }


   }



   num_population=j1;



   for(int  ii=0;ii<=num_population;ii++)
   repetition(population,ii,num_population);

for(int  i=0;i<num_population;i++)
feasibility(i);


  for  (i=0;  i<num_population;  i++)  {
   int  y=0;
            for(k=0;k<flows;k++){
            j=0;sh=0;
                    while(population[i][k][j]!=-1 && sh==0){
                    if(verify_capacity( population[i][k][j],
population[i][k][j+1],k))
                            {
                            j++ ;
                            }else{
                            remove_path_2(i);
                            sh=1;
                            k=flows;
                            i--;
                            restart_capacity();
                            }
                    }

            }
            j1++;
```

```
   }


   /*
i=0;
  while(population[i][0][0]!=-1 && population[i][0][1]!=0)
  {i++;
  }
 num_population=i;   */
 num_population_initial  =num_population;
}
//end initial_population()--------------------------------
//SetFit()-----------------------------------------------
void SetFit(int p,int i,  int s){

    strength[i][p]=s;
}
//end SetFit()------------------------------------------
//GetFit()----------------------------------------------
float GetFit(int p,int i){

    return strength[i][p];
}
//end GetFit()-----------------------------------------
//dominatesOrEqual--------------------------------------
bool dominatesOrEqual(int i,int j,int
cam1[MAX_POPULATION][MAX_FLOWS][MAX_NODES],int
cam2[MAX_POPULATION][MAX_FLOWS][MAX_NODES]){

  int k=0,l,state1,state2,h;
  int win,loss,tie;

          int fit[2][2];
          fit[0][0]=0;
          fit[0][1]=0;
          fit[0][2]=0;
          fit[1][0]=0;
          fit[1][1]=0;
          fit[1][2]=0;
```

```
           for(h=0;h<flows;h++){
           while (cam1[i][h][k+1] != -1) {
           fit[0][0] = fit[0][0] + 1;//hop count
 fit[0][1] = fit[0][1] +
delay[cam1[i][h][k]][cam1[i][h][k+1]];//delay
           fit[0][2] = fit[0][2] + demand[h];
 k++;
 }
           }
 k=0;

           for(h=0;h<flows;h++){
           while (cam2[j][h][k+1] != -1) {
           fit[1][0] = fit[1][0]+1;//hop count
 fit[1][1] = fit[1][1] +
delay[cam2[j][h][k]][cam2[j][h][k+1]];//delay
           fit[1][2] = fit[1][2] + demand[h];
 k++;
 }
           }
     win   = 0;
     loss = 0;
     tie = 0;
     if (fit[0][0] < fit[1][0])
         win++;
     else if (fit[0][0] == fit[1][0])
             tie++;
           else
               loss++;

     if (fit[0][1] < fit[1][1])
         win++;
     else if (fit[0][1] == fit[1][1])
             tie++;
           else
               loss++;

                 if (fit[0][2] < fit[1][2])
         win++;
     else if (fit[0][2] == fit[1][2])
```

```
                    tie++;
            else
                loss++;



                if(win>loss)
                return true;
                else
                  if(win==loss)
                        return true;
                    else
                        return false;


}
//end dominatesOrEqual----------------------------------
//calc. Strengths and fitness for non-dominated population-
------------------
void   calcStrengths()
{

   int    i, j;
   int    domCount,pop_count;

   for (i = new_num_population1 - 1; i >= 0; i--) {

      domCount = 0;
      pop_count=0;
      for (j = num_population - 1; j >= 0; j--){

         if(population[j][0][0]!=-1){
         pop_count++;
         if (dominatesOrEqual(i,j,new_population1,
population)){
  domCount++;
         }
         }
      }
      if(domCount==0)pop_count=domCount=1;
```

```
        SetFit(0,i, 1/(double(domCount) / double(pop_count +
1))); //num_population
   }
}
//end calcStrengths()-------------------------------------
//calc.Fitness for dominated population-------------------
void  calcFitness()
{

   int       i, j;
   double    fitness1;

   for (i = num_population - 1; i >= 0; i--) {

      fitness1 = 0.0;
    if(population[i][0][0]!=-1){
    for (j = new_num_population1 - 1; j >= 0; j--) {

      if (dominatesOrEqual(j,i,new_population1,population))
 fitness1 += GetFit(0,j);
      }
      SetFit(1,i, 1.0 + fitness1);
      }
   }
}
//end calcFitness()--------------------------------------
//Distance calculate-------------------------------------
double  IndDistance(int i,int j,int
cam1[MAX_POPULATION][MAX_FLOWS][MAX_NODES])
{
double distance;
distance=0;

           int fit[2][2],k=0,h;
           fit[0][0]=0;
           fit[0][1]=0;
           fit[0][2]=0;
           fit[1][0]=0;
           fit[1][1]=0;
           fit[1][2]=0;
```

```
         for(h=0;h<flows;h++){
         k=0;
         while (cam1[i][h][k+1] != -1) {
         fit[0][0] = fit[0][0]+1;//hop count
  fit[0][1] = fit[0][1] +
delay[cam1[i][h][k]][cam1[i][h][k+1]];//delay
         fit[0][2] = fit[0][2] + demand[h];
  k++;
 }
            }


         for(h=0;h<flows;h++){
         k=0;
         while (cam1[j][h][k+1] != -1) {
         fit[1][0] = fit[1][0]+1;//hop count
  fit[1][1] = fit[1][1] +
delay[cam1[j][h][k]][cam1[j][h][k+1]];//delay
         fit[1][2] = fit[1][2] + demand[h];
  k++;
 }
            }
   distance = sqrt(
                     (pow(fit[0][0] - fit[1][0], 2)) +
                     (pow(fit[0][1] - fit[1][1], 2)) +
                     (pow(fit[0][2] - fit[1][2], 2))
                      );
   return distance;
}
//End distance----------------------------------------------
//Cluster distance------------------------------------------
double  ClusterDistance(int  ic1, int  ic2)
{
   double  sum = 0;
   int  numOfPairs = 0;


   int ik1=0;
   while(clusters[ic1][ik1]!=-1)//clusters[ic1][ik1]
   {
```

```
            int ik2=0;
            while(clusters[ic2][ik2]!=-1){
             numOfPairs++;
             sum += IndDistance(ic1,ic2,new_population);
              ik2++;
            }
            ik1++;
   }

if(sum<0)  sum=sum*-1;

return (sum / double(numOfPairs));

}
//End clusters distance---------------------------------
//Join Clusters----------------------------------------
void  JoinClusters(int   cluster1, int   cluster2, int&
numOfClusters)
{

int k=0;
while(clusters[cluster1][k]!=-1)
{
  k++;
}

int kk=0;
while(clusters[cluster2][kk]!=-1)
{clusters[cluster1][k]=clusters[cluster2][kk];
k++;kk++;
}

k=0;
while(clusters[cluster2][k]!=-1)
{clusters[cluster2][k]=-1;k++;}

numOfClusters--;
}
//End Join Clusters------------------------------------
//Find Cluster Centroid--------------------------------
```

```
int   ClusterCentroid(int cluster)
{

        int  cluster_value[200][2],k=0,prom1=0,prom2=0;

        for(int  i=0;i<100;i++)
        {cluster_value[i][0]=0;cluster_value[i][1]=0;}

        int num_cam=0;
        while(clusters[cluster][k]!=-1){
                int cl=clusters[cluster][k];
                int kk=0;
                for(int  h=0;h<d;h++){
                while (new_population[cl][h][kk+1] != -1) {
                        cluster_value[cluster][0]  =
cluster_value[cluster][0]+1;//hop count
                cluster_value[cluster][1]  = cluster_value
[cluster][1]+delay[new_population[cl][h][kk]][new_
population[cl][h][kk+1]];//delay
                kk++;
                }

                }

            prom1=cluster_value[cluster][0];
            prom2=cluster_value[cluster][1];
            num_cam++;
            k++;

        }
        prom1=prom1/num_cam;
        prom2=prom2/num_cam;

    for(int  i=0;i<100;i++)
    {cluster_value[i][0]=0;cluster_value[i][1]=0;}
    int  min_dist=1000,selected=0;
    float  dist=0;
    k=0;

    while(clusters[cluster][k]!=-1){
```

```
  kk=0;
  int cl=clusters[cluster][k];
    cluster_value[cluster][0]=0;
    cluster_value[cluster][1]=0;

    for(int h=0;h<flows;h++){
    while (new_population[cl][h][kk+1] != -1) {
                    cluster_value[cluster][0] =
cluster_value[cluster][0]+1;//hop count
            cluster_value[cluster][1] = cluster_value
[cluster][1] +
delay[new_population[cl][h][kk]][new_population[cl][h][kk
+1]];//delay
            kk++;
    }
    }

      dist=sqrt(pow(cluster_value[cluster][0]-
prom1,2)+pow(cluster_value[cluster][1]-prom2,2));

      if(dist<min_dist)
      {
         selected=clusters[cluster][k];
         min_dist=dist;
      }
    k++;

  }

  clusters[cluster][0]=selected;
  k=1;
  while(clusters[cluster][k]!=-1){
       clusters[cluster][k]=-1;
       k++;
  }

  for(int h=0;h<flows;h++){
  k=0;
  while(new_population[clusters[cluster][0]][h][k]!=-1){
new_population1[new_num_population1][h][k]=new_population
[clusters[cluster][0]][h][k];
```

```
    k++;
    }
    new_population1[new_num_population1][h][k]=-1;
    }
    new_num_population1++;
}
//End Cluster Centroid------------------------------------
//Reduce Non-Dominated------------------------------------
void  clustering(void)
{

int i;
for (i = 0 ; i < new_num_population; i++){
clusters[i][0]=i;
}

    int population_clusters=new_num_population;
    while(population_clusters>maxNondominated){

        float minDist=100;
        int join1,join2;
        int  numOfClusters = new_num_population;

        for (int  c = 0; c < numOfClusters; c++) {
        for (int  dd = c + 1; dd < numOfClusters; dd++) {

          if(clusters[c][0]!=-1 && clusters[dd][0]!=-1)
          {
                  double dist=ClusterDistance(c,dd);
//clusters[c][0],clusters[d][0]);

            if (dist < minDist) {
              minDist = dist;
               join1 = c;
               join2 = dd;
                      }}
          }
          }

    JoinClusters(join1, join2, population_clusters);
```

```
      }
      int  c=0,ind=0;
      while(c<population_clusters)
      {

      if(clusters[ind][0]!=-1)
      {
      ClusterCentroid(ind);
      c++;
      }
      ind++;
      }
}
//End Reduce Non-Dominated-----------------------------
//Bubble_Sort-------------------------------------------
void Bubble_Sort()
{
   int   i,j;
   float temp1=0,temp2=0;

      for(i=0;i<new_num_population1;i++){
          for(j=i+1;j<new_num_population1;j++){

                  if(Prob[i][0]>Prob[j][0])
                  {
                     temp1=Prob[i][0];
                     temp2=Prob[i][1];
                     Prob[i][0]=Prob[j][0];
                     Prob[i][1]=Prob[j][1];
                     Prob[j][0]=temp1;
                     Prob[j][1]=temp2;

                  }
              }
          }

      for(i=0;i<num_prob;i++){
          for(j=i+1;j<num_prob;j++){

                  if(Prob[i][2]>Prob[j][2])
```

```
                    {
                        temp1=Prob[i][2];
                        temp2=Prob[i][3];
                        Prob[i][2]=Prob[j][2];
                        Prob[i][3]=Prob[j][3];
                        Prob[j][2]=temp1;
                        Prob[j][3]=temp2;
                    }
                }
            }
}


//End Bubble_Sort---------------------------------------
//Join_Prob---------------------------------------------
void Join_Prob()
{
    int   temp1=0,temp2=0,i,j,ii=0;

    Jprob[0][0]=0;
    Ci[0][7]=0;
    Ci[0][8]=0;


        for(i=0;i<num_prob;i++){

            if(i==0){
            Jprob[i][0]=Prob[i][2];
            }else{
            Jprob[i][0]=Jprob[i-1][0]+Prob[i][2];
            }
            Jprob[i][1]=1;
            Jprob[i][2]=Prob[i][3];
        }
        j=i;
        for(i=0;i<new_num_population1;i++){
        Jprob[j][0]=Jprob[j-1][0]+Prob[i][0];
        Jprob[j][1]=0;
        Jprob[j][2]=Prob[i][1];
        j++;
```

```
          }

}

//End Join_Prob-----------------------------------------
//Set_new_population------------------------------------
void Set_new_population(int p,int i)
{
int  k,h;

if(p==0){

   for(h=0;h<flows;h++){
   k=0;
   while(new_population1[i][h][k]!=-1)
   {

new_population[new_num_population][h][k]=new_population1[
i][h][k];
    k++;
   }
   new_population[new_num_population][h][k]=-1;
   }
   new_num_population++;
}else{

for(h=0;h<flows;h++){
k=0;
   while(population[i][h][k]!=-1)
   {
   new_population[new_num_population][h][k]=
population[i][h][k];
    k++;
   }
   new_population[new_num_population][h][k]=-1;
   }
   new_num_population++;
}

}
```

```
//Set_new_population------------------------------------
//Roulette_Selection------------------------------------
void    Roulette_Selection()
{

        float  num_random,sum_fitness=0,sum_ci=0,prob[100][3];
        int  i,j,k;
        for(i=0;i<MAX_POPULATION;i++)
        {
        for(k=0;k<MAX_FLOWS;k++){
            for(j=0;j<MAX_NODES;j++)
              {
                new_population[i][k][j]=-1;
              }
        }
        }

        new_num_population=0;

        for(i=0;i<new_num_population1;i++){
        sum_fitness=sum_fitness+GetFit(0,i);
        Ci[i][0]=0;
        }

        for(i=0;i<num_population;i++){
        if(population[i][0][0]!=-1){
        sum_fitness=sum_fitness+GetFit(1,i);
        Ci[i][1]=0;
        }
        }

        for(i=0;i<new_num_population1;i++){
        Ci[i][0]=1-double(GetFit(0,i)/sum_fitness);
        sum_ci=sum_ci+Ci[i][0];
        }

        for(i=0;i<num_population;i++){
        if(population[i][0][0]!=-1){
        Ci[i][1]=1-double(GetFit(1,i)/sum_fitness);
        sum_ci=sum_ci+Ci[i][1];
```

```
        }
        }

    for(i=0;i<new_num_population1;i++){
    Prob[i][0]=Ci[i][0]/sum_ci;
    Prob[i][1]=i;


        }

    j=0;

    for(i=0;i<num_population;i++){
    if(population[i][0][0]!=-1){
    Prob[j][2]=Ci[i][1]/sum_ci;
    Prob[j][3]=i;
    j++;


        }
        }

    num_prob=j;

    Bubble_Sort();
    Join_Prob();

    new_num_population=0;
    for(i=0;i<(num_prob+new_num_population1)/2;i++){
    num_random=rand()/32300.0;
    int  sw=0;
    j=0;
    while((j<num_population+new_num_population1-
1)&&(sw==0)){
      if(Jprob[0][0]>num_random){
      Set_new_population(Jprob[0][1],Jprob[0][2]);
      sw=1;
      }else{

if((Jprob[j+1][0]>=num_random)&&(Jprob[j][0]<num_random))
{
            Set_new_population(Jprob[j+1][1],Jprob[j+1][2]);
```

```
            sw=1;
        }
    j++;
        }
        }
        }


}
//end Roulette_Selection----------------------------------
void select_non_dominated()   {

            for (int i=0; i<MAX_POPULATION; i++) {
            for(int h=0;h<MAX_FLOWS;h++){
    for (int j=0; j<MAX_NODES; j++) {
        new_population[i][h][j] = -1;
    }
                }
            }
            int i,j,sw,k,l,h,state1,state2;
  int win,loss,tie;
            if (num_population == 1) {
        k=0;
        for (l=0; l < flows; l++) {
            while (population[0][l][k] != -1) {
                new_population[0][l][k] = popula-
tion[0][l][k];
                                population[0][l][k]=-1;//ojo
                k++;
            }
            new_population[0][l][k] = population[0][l][k];
                                population[0][l][k]=-1;//ojo
            k++;
        }
        new_num_population++;
    }else{

  new_num_population = 0;
  for(i=0; i<num_population; i++) {
                    j=0;
                    sw=0;
```

```
              if(population[i][0][0]!=-1)
              {

                  while((j<num_population)&&(sw==0)){
                  win    = 0;
loss = 0;
tie = 0;

                  if (fitness[i][0] < fitness[j][0])
    win++;
else if (fitness[i][0] == fitness[j][0])
        tie++;
     else
         loss++;

if (fitness[i][1] < fitness[j][1])
    win++;
else if (fitness[i][1] == fitness[j][1])
        tie++;
     else
         loss++;

                  if (fitness[i][2] < fitness[j][2])
    win++;
else if (fitness[i][2] == fitness[j][2])
        tie++;
     else
         loss++;

                  if (win > 0) {
    if (loss == 0) {
        state1 = NONDOMINATED;
        state2 = DOMINATED;
    }
    else {
        state1 = BOTH_NONDOMINATED;
        state2 = BOTH_NONDOMINATED;
    }
}
else {
```

```
        if (loss > 0) {
                                    sw=1;
            state1 = DOMINATED;
            state2 = NONDOMINATED;
        }
        else {
            state1 = BOTH_NONDOMINATED;
            state2 = BOTH_NONDOMINATED;
        }
    }

    if (tie == 6) {
        state1 = NONDOMINATED;
        state2 = DOMINATED;
    }
                    j++;
                }


                }


                        if(sw==0){
                            h=0;
                            while(h<flows){
                            k=0;
                    while (population[i][h][k]
!= -1) {
            new_population[new_num_population][h][k]
= population[i][h][k];
                            population[i][h][k]=-1;
                k++;
            }
            new_population[new_num_population][h][k] =
population[i][h][k];

                            population[i][h][k]=-1;
                k++;
                            h++;
                }
        new_num_population++;
```

```
                                /*  int jjj=i;
                                   while(jjj<num_population-1){
                                   for(h=0;h<d;h++){
                                          k=0;
                                 while(population[jjj+1][h][k]!=-1){
                                          population[jjj][h][k]=
population[jjj+1][h][k];
                                          k++;
                                          }
                                     population[jjj][h][k]=-1;
                                   }

                                   jjj++;
                                   }
                                   num_population--;
                                         */
                                   }

       }


          for(i=0;  i<num_population;  i++){
                     if(population[i][0][0]==-1)
                     {
                     j=i;
                                 while(j<num_population){
                                 for(h=0;h<flows;h++){
                                        k=0;
                                  while(population[j+1][h][k]!=-1){
                                          population[j][h][k]=
population[j+1][h][k];
                                          k++;
                                          }
                                      population[j][h][k]=-1;
                                 }
                                 j++;
                                 }
                                 num_population--;
                     }
              }
```

```
            }

}

//SPEA------------------------------------------------
------------------
void SPEA()
{
int i,k,h;
   select_non_dominated();
  if (new_num_population > maxNondominated){
  new_num_population1=0;
  clustering();
  }else{
  new_num_population1=0;
  for(i=0;i<new_num_population;i++){
          for(h=0;h<flows;h++){
          k=0;
          while(new_population[i][h][k]!=-1){

new_population1[new_num_population1][h][k]=new_population
[i][h][k];
          k++;
          }
          new_population1[new_num_population1][h][k]=-1;
          }
          new_num_population1++;
  }
  }
  calcStrengths();
  calcFitness();
}
//End SPEA----------------------------------------------
//Functions Calculate ----------------------------------
void Functions_calculate(int
pop[MAX_POPULATION][MAX_FLOWS][MAX_NODES],int num_pop) {

 int i,j,k,max,min,max2,min2;
          max=0;
          min=1000;
          max2=0;
```

```
         min2=1000;
 float alpha;

 for (i=0; i<num_pop; i++) {
     for (j=0; j<3; j++) {
         fitness[i][j] = 0.0;
     }

 }
 for (i=0; i<num_pop; i++) {
     k=0;
     for (j=0; j < flows; j++) {
                 k=0;
         while (pop[i][j][k+1] != -1) {
             fitness[i][0] = fitness[i][0] + 1;//hop
count
             fitness[i][1] = fitness[i][1] +
delay[pop[i][j][k]][pop[i][j][k+1]];//delay
                                     fitness[i][2] =
fitness[i][2] + demand[j];
             k++;
         }

         k++;
     }
 }


}
//end Functions calculate------------------------------
//end restriction------------------------------------------
void restriction(int paths-tmp
[MAX_POPULATION][MAX_FLOWS][MAX_NODES],int num){
for(int i=0;i<num;i++){
for(int h=0;h<flows;h++){
         int j=0,sw=0;
         while((paths-tmp[i][h][j]!=-1)&&(sw==0)){


             if(capacity[paths-tmp[i][h][j]][paths-
tmp[i][h][j+1]]<demand[h])
```

```
                    {
                        if(capacity[paths-tmp[i][h][j]][paths-
tmp[i][h][j+1]]!=0){


                            for(int ii=i;ii<=num;ii++){

                            for(int hh=0;hh<flows;hh++){
                              int jj=0;

                                while(paths-tmp[ii][hh][jj]!=-1)
                                  {
                                    paths-tmp[ii][hh][jj]=paths-
tmp[ii+1][hh][jj];

                                  jj++;

                                  }
                                }

                            }

                        sw=1;
                        }
                      }
                      j++;  }
}

      }



}
//end restriction-------------------------------------------
//cross-over pointer-----------------------------------------
int crossover_position(int i,int des){
int sw=0,j=0;
while(sw==0){
        if((new_population[i][des][j]==
-1)||(new_population[i+1][des][j]==-1)){
            sw=1;
            }
        j++;
}
```

```
return j-2;
}
//end cross-over pointer--------------------------------
//verify_path------------------------------------------
bool verify_path(int
pop[MAX_POPULATION][MAX_FLOWS][MAX_NODES], int indice){
    int sw;
    sw=0;
    for(int h=0;h<flows;h++){
    int j=0;

            while(((pop[indice][h][j+1]!=-1)&&(sw==0))
            {
                if(capac-
ity[pop[indice][h][j]][pop[indice][h][j+1]]==0)
                {sw=1;h=flows;}
            j++;
            }
    }
    if(sw==0)
    return false;
    else
    return true;


}
//end verify_path--------------------------------------
//remove_path_3----------------------------------------
void remove_path_3(int indice, int
pop[MAX_POPULATION][MAX_FLOWS][MAX_NODES],int size){

        int i=0,j,h;
        for(h=0;h<flows;h++){
        for(i=indice;i<MAX_NODES;i++)
        {pop[indice][h][i]=-1;}
        }

        for(i=indice;i<size-1;i++){
        for(h=0;h<flows;h++){
            j=0;
            while(pop[i+1][h][j]!=-1)
```

```
                  {
                  pop[i][h][j]=pop[i+1][h][j];
                  j++;
                  }
               for(int k=j;k<MAX_NODES;k++){pop[i][h][k]=-1;}
            }


         }


}
//end remove path----------------------------------------
//Repeated_chromosomes ----------------------------------
void repeated_chromosomes(){

for (int i=0; i<num_population-1; i++) {
        for(int h=0;h<flows;h++){
          int j=0;
   while(population[i][h][j]!=-1){


                 for (int ii=0; ii<num_population; ii++) {

                         int jj=0;
                     while(population[ii][h][jj]!=-1){



                         }

                   }



               }
            }
      }


}
//end Repeated_chromosomes ------------------------------
//Repeated_nodes ----------------------------------------
```

```
bool Repeated_nodes(int
pop[MAX_POPULATION][MAX_FLOWS][MAX_NODES], int indice){
int i=0,j=0,sw=0,h;

for(h=0;h<flows;h++){
while(pop[indice][h][i+1]!=-1)
{
    j=0;
    while(pop[indice][h][j]!=-1){

            if(i!=j){

                if(pop[indice][h][i]==pop[indice][h][j]){
                    sw=1;
                    h=d;
                }
            }
             j++;
            }
        i++;
}
}


if(sw==0)
return false;
else
return true;


}
//End Repeated_nodes -----------------------------------
//cross over -------------------------------------------
// Figure 4.33
void cross_over() {

for (int i=0; i<MAX_POPULATION; i++) {
        for(int h=0;h<flows;h++){
    for (int j=0; j<MAX_POPULATION; j++) {
        new_population2[i][h][j] = -1;
```

```
        }
            }
        }


    int i,l,k,k1,k2,k3,cross_over_position,dest_rand;
    float num_random;
    new_num_population2 = 0;


            if(new_num_population == 1){


                    for(int h=0;h<flows;h++){
                    k3=0;
                    while(new_population[0][h][k3]!= -1){

new_population2[0][h][k3]=new_population[0][h][k3];
                    k3++;
                    }

new_population2[0][h][k3]=new_population[0][h][k3];
                    }
            }else{

            for(i=0; i<new_num_population-1; i++){
            num_random = rand()/32300.0;

            if (num_random <= CROSS_OVER_PROBABILITY) {
            for(int h=0;h<flows;h++){

                    cross_over_position=ran-
dom(crossover_position(i,h));

                    if(cross_over_position==0){
                    cross_over_position=1;
                    }

                    k=0;k1=0;k2=0;
                    while(k<cross_over_position){
                new_population2[new_num_population2][h][k]
= new_population[i][h][k];
                new_population2[new_num_population2+1][h][k]
= new_population[i+1][h][k];
```

```
                        k++;
                        }
                        k1=k;
                        k2=k;
                      while(new_population[i+1][h][k1]!= -1){

new_population2[new_num_population2][h][k1]=new_population
[i+1][h][k1];
                        k1++;
                        }

new_population2[new_num_population2][h][k1]=new_population
[i+1][h][k1];

                      while(new_population[i][h][k2]!= -1){

new_population2[new_num_population2+1][h][k2]=new_population
[i][h][k2];
                        k2++;
                        }

new_population2[new_num_population2+1][h][k2]=new_population
[i][h][k2];

                }
              new_num_population2=new_num_population2+2;
                }
                }

                }


        for(i=0; i<new_num_population; i++){
                    for(int h=0;h<flows;h++){
                    int k4=0;
                    while(new_population[i][h][k4]!= -1){

new_population2[new_num_population2][h][k4]=new_population
[i][h][k4];
                        k4++;
                        }
```

```
new_population2[new_num_population2][h][k4]=new_population
[i][h][k4];
                        k4++;
                        }
                    new_num_population2++;
            }


    for(k=0;k<new_num_population2;k++)
     {
    if(Repeated_nodes(new_population2,k))        // repeated
remove
    {remove_path_3(k,new_population2,new_num_population2);
        new_num_population2--;
       k--;
       }else{
            if(verify_path(new_population2,k))      // Path
Feasible

{remove_path_3(k,new_population2,new_num_population2);
                new_num_population2--;
                k--;
                }
        }

     }


}
//end cross_over()--------------------------------------
//mutation position------------------------------------
int m_position(int i,int des){
int sw=0,j=0;
while(sw==0){
        if(new_population[i][des][j]==-1){
                sw=1;
            }
        j++;
    }
return j-2;
```

```
}
//end_mutation_position-------------------------------
//mutation---------------------------------------------
// Figure 4.35
void mutation() {

for (int i=0; i<MAX_POPULATION; i++) {
        for(int h=0;h<flows;h++){
    for (int j=0; j<MAX_POPULATION; j++) {
        new_population3[i][h][j] = -1;
    }
        }
    }
        int
i,l,k,k1,k2,u,i2,mutation_position,p,mutation_rand,h;
 float num_random;
        int paths_tmp[MAX_NODES];

 new_num_population3 = 0;

        for(i=0; i<new_num_population2-1; i++){

        num_random = rand()/32300.0;

                if (num_random <= MUTATION_PROBABILITY){

                        mutation_rand=random(flows);

                        mutation_position=ran-
dom(m_position(i,mutation_rand));

                        if(mutation_position<=1){
                        mutation_position=2;
                        }
                        k=0;k1=0;

                                while(k<mutation_position){
new_population3[new_num_population3][mutation_rand][k] =
new_population2[i][mutation_rand][k];
                                k++;
```

```
                                }


if(bfs(new_population2[i][mutation_rand][k],dest[0])){

                               paths_tmp[0] = dest[0];
               for (u=dest[0]; pred[u]>=0; u=pred[u]) {
                                         k1++;
                    paths_tmp[k1] = pred[u];
                                visited[pred[u]][u] = 0;
               }



new_population3[new_num_population3][mutation_rand][k] =
pred[u];

               for (int i1 = k1; i1>=0; i1--) {

new_population3[new_num_population3][mutation_rand][k]=pa
ths_tmp[i1];
               k++;
               }

new_population3[new_num_population3][mutation_rand][k]=
-1;

                               for(h=0;h<flows;h++){

                                 if(h!=mutation_rand)
                                 {
                                     k=0;
                         while(new_population2[i][h][k]!=
-1){

new_population3[new_num_population3][h][k]=new_population
2[i][h][k];

                                     k++;
                                 }
new_population3[new_num_population3][h][k]=-1;
                                 }
                             }
```

```
                                         new_num_population3++;
                                       }
                         }
               }


         for(k=0;k<new_num_population3;k++)
       {
         if(Repeated_nodes(new_population3,k))
      {remove_path_3(k,new_population3,new_num_population3);
           new_num_population3--;
          k--;
         }else{
              if(verify_path(new_population3,k))
{remove_path_3(k,new_population3,new_num_population3);
                  new_num_population3--;
                  k--;
                }
         }
       }




}
//end mutation()-----------------------------------------
void Clear_cluster(){

   for(int i=0;i<100;i++){
   for(int j=0;j<500;j++){
   clusters[i][j]=-1;
   }
   }
}
//UpDate new populations---------------------------------
-------------------
void update_new_population() {

  int i,j,k,l;
```

```
        int num_temp_population=0;

        for (i=0; i<MAX_POPULATION; i++) {
        for(l=0;l<flows;l++){
    for (j=0; j<MAX_POPULATION; j++) {
                temp_population[i][l][j] = -1;
                }
                }
}


        int ii=0;
  for (i=0; i<num_population; i++) {

                if(population[i][0][0]!=-1){
                  for(l=0;l<flows;l++){
                  j=0;
    while(population[i][l][j]!=-1){

                        temp_population[ii][l][j]=
population[i][l][j];

                        j++;

                }
                temp_population[ii][l][j]=-1;
                }
                ii++;
                }
}

        num_temp_population=ii;

        for (i=0; i<MAX_POPULATION; i++) {
        for(l=0;l<flows;l++){
        for (j=0; j<MAX_POPULATION; j++)
 {population[i][l][j]=-1;}}}

        num_population = 0;
        for(i=0;i<new_num_population1;i++){
```

```
            if(new_population1[i][0][0]!=-1)
            {
            for(l=0;l<flows;l++){
            j=0;
    while(new_population1[i][l][j]!=-1){
                     popula-
tion[num_population][l][j]=new_population1[i][l][j];
                     j++;
                     }
                     population[num_population][l][j]=-1;
            }
                     num_population++;
            }
                     }


        for(i=0;  i<new_num_population2;  i++) {
                for(l=0;l<flows;l++){
                k=0;
    while  (new_population2[i][l][k]  != -1)  {

    population[num_population][l][k]  =
new_population2[i][l][k];
    k++;
    }
    population[num_population][l][k]  =
new_population2[i][l][k];
    k++;
                }
                num_population++;

  }

  for(i=0;  i<new_num_population3;  i++)  {
                for(l=0;l<flows;l++){
                k=0;
    while  (new_population3[i][l][k]  != -1)  {
    population[num_population][l][k]  =
new_population3[i][l][k];
    k++;
    }
```

```
     population[num_population][l][k]  =
new_population3[i][l][k];
     k++;
                        }
                    num_population++;

  }




     for (i=0; i<MAX_POPULATION; i++) {
     for(l=0;l<flows;l++){
   for (j=0; j<MAX_POPULATION; j++) {
                    new_population[i][l][j]  = -1;
                    new_population1[i][l][j]  = -1;
                    new_population2[i][l][j]  = -1;
                    new_population3[i][l][j]  = -1;
                    temp_population[i][l][j]  = -1;
                    }
                    }
  }

     for(int ii=0;ii<=num_population;ii++)
   repetition(population,ii,num_population);


     Clear_cluster();
}
//end update_new_population()----------------------------
//population_view()------------------------------------
void population_view(int
population_v[MAX_POPULATION][MAX_FLOWS][MAX_NODES], int
size_population,int count_generations ,String title) {
//char *title
 int i,k,l,g,h;

 printf("\n\n%s\n\n",title);
         printf("GENERATION \tPOPULATION\n");
 for(i=0; i < size_population; i++) {
     k=0;
                    printf("%d \t\t",count_generations);
```

```
                          for(l=0;l<flows;l++){
                          k=0;
           while (population_v[i][l][k] != -1) {
              printf("%d,",population_v[i][l][k]);
                                    k++;
           }
                          printf("\n\t\t");
                          }
                          printf("\n");
           k++;
 }
}
//end population_view()----------------------------------
//FITNESS VIEW------------------------------------------
void fitness_view(int tam,int generation) {
 int i;

 printf("\FUNCTION VALUES\n\n");
        printf("CHROMOSOME \t FUNCTIONS \t GENERATION\n");
 for(i=0; i < tam; i++) {

     printf("Chromosome %d \t Hop Count \t %f\n",i,fit-
ness[i][0]);
     printf("Chromosome %d \t Delay      \t %f\n",i,fit-
ness[i][1]);
                    printf("Chromosome %d \t Bandwidth \t
%f\n",i,fitness[i][2]);
          }
 printf("\n");

}
//end fitness_view()------------------------------------
//------------------------------------------------------
//------------------------------------------------------
void B_INPUTFILE()
{
int a,b,c,h,i,j,l;
     FILE* input = fopen("mf_MOEA.in","r");
     // read number of nodes, edges and sinks
     fscanf(input,"%d %d %d %d\n",&n,&e,&d,&flows);
```

```
    //printf("%d %d\n",n,e);
    // initialize empty capacity matrix
    for (i=0; i<n; i++) {
    num_paths_by_destination[i][0] = 0;
    num_paths_by_destination[i][1] = 0;
for (j=0; j<n; j++) {
    capacity[i][j] = 0;
    delay[i][j] = 0;
    visited[i][j] = 0;
}

    }
    for (i=0; i<MAX_POPULATION; i++) {
    for(l=0;l<MAX_FLOWS;l++){
for (j=0; j<MAX_NODES; j++) {
    population[i][l][j] = 0;
}

        }
    }

    // read sink nodes
printf("SINKS NODES:\t");
for (i=0; i<d; i++) {
    fscanf(input,"%d",&dest[i]);
        printf("%d , ",dest[i]);

                    target=dest[i];
}
// read demand of the multicast flow (kbps)
        printf("\nNUM. OF FLOWS:\t %d",flows);
printf("\nDEMAND FLOW:\t");
        for (i=0; i<flows; i++) {
fscanf(input,"%d\n",&demand[i]);printf("%d Kbps ,
",demand[i]);
        }

    // read edge capacities
printf("\n\nEDGE CAPACITIES\n");

    for (i=0; i<e; i++) {
    fscanf(input,"%d %d %d %d",&a,&b,&c,&h);
```

```
     printf("%d %d %d %d\n",a,b,c,h);

                capacity[a][b] = c;
     delay[a][b] = h;
     visited[a][b] = 1;
     }

     int r=0;
     r=visited[8][12];

     fclose(input);
}
//------------------------------------------------------
void OUTFILE(int size_population,int
population_v[MAX_POPULATION][MAX_FLOWS][MAX_NODES]){
FILE* output = fopen("OUTPUT_FILE.txt","w");
     // read number of nodes, edges and sinks
     fprintf(output,"%s\n\n","FINAL POPULATION");
     fprintf(output,"GENERATION \tPOPULATION\n\n");
          int i,k,l;
  for(i=0; i < size_population; i++) {

                                fprintf(output,"%d \t\
t",count_generations);
                                for(l=0;l<flows;l++){
                                k=0;
          while (population_v[i][l][k] != -1) {
              fprintf(output,"%d,",population_v[i][l][k]);
                                     k++;
          }
                                fprintf(output,"\n\t\t");
                                }
                                fprintf(output,"\n\n");
          k++;
  }

          fprintf(output,"\nFUNCTION VALUES\n\n");
          fprintf(output,"CHROMOSOME \t FUNCTIONS \t
GENERATION\n\n");
  for(i=0; i < size_population; i++) {
```

```
      fprintf(output,"Chromosome %d \t Hop Count\t %f\
n",i,fitness[i][0]);
      fprintf(output,"Chromosome %d \t Delay    \t %f\
n",i,fitness[i][1]);
                  fprintf(output,"Chromosome %d \t
Bandwidth\t %f\n",i,fitness[i][2]);
         }

 fprintf(output,"\n");
         fclose(output);


}
//----------------------------------------------------------
/*void sort_best(){
int i,j,k,salt1,salt2,temp;
for(i=0;i<num_population_initial   ;i++){
         k=0;
         salt1=0;
         while(population_branch[i][k]!=-1){
         salt1++;
         k++;
         }
for(j=i+1;j<num_population_initial;j++){
         k=0;
         salt2=0;
         while(population_branch[j][k]!=-1){
         salt2++;
         k++;
         }

         if(salt2<salt1){
         k=0;
         while(k<=salt1){
         temp=population_branch[i][k];
         population_branch[i][k]=population_branch[j][k];
         population_branch[j][k]=temp;

         k++;
         }
         salt1=salt2;
```

```
               population_branch[j][k]=-1;
               population_branch[i][k]=-1;
               }
      }
      }
      }
//-----------------------------------------------------------
-------------------
void Search_trees(){
   int h,k,i,npi=1000;

      for(h=0;h<d;h++){
           target=dest[h];
           Search_Paths(0);    // Source node : node 0

           sort_best();

           for(i=0;i<num_population_initial;i++){
                   k=0;
                   while(population_branch[i][k]!=-1)
                   {
               population[i][h][k]=population_branch[i][k];
                   k++;
                   }
                   population[i][h][k]=-1;
           }

           if(npi>num_population_initial)
           npi=num_population_initial;

         num_population_initial=0;
         clear_branch();
       }

      num_population_initial=npi;
}          */
//-----------------------------------------------------------

int main ()
{
```

```
        B_INPUTFILE();
        maxNondominated = MAX_NONDOMINATED_SOLUTIONS;
        int    nondominatedSetUpperBound = num_population
+ maxNondominated;

        time1 = clock();

        Search_Paths(0);
        initial_sort();
 initial_population();

        population_view(popula-
tion,num_population_initial,0,"INITIAL POPULATION");
        num_population=num_population_initial;
 count_generations = 0;

 while (count_generations < MAX_GENERATIONS) {
     //printf("\n\nGeneration: %d\n\
n",count_generations+1);
     Functions_calculate(population,num_population);
     //fitness_view(tringGrid10,count_generations);
                 Clear_cluster();
                 SPEA();
                 Roulette_Selection();
     cross_over();

population_view(new_population2,new_num_population2,count
_generations,"POPULATION AFTER CROSSOVER");
     mutation();

population_view(new_population3,new_num_population3,count
_generations,"POPULATION AFTER MUTATION   ");
     update_new_population();
                 population_view(popula-
tion,num_population,count_generations,"NEW POPULATION");
     count_generations++;
 }


        for(int ii=0;ii<num_population;ii++)
        repetition(population,ii,num_population);
```

```
        population_view(popula-
tion,num_population,count_generations,"NEW POPULATION");

        Functions_calculate(population,num_population);
        select_non_dominated();

population_view(new_population,new_num_population,4,"FINAL
POPULATION");

Functions_calculate(new_population,new_num_population);
        fitness_view(new_num_population,count_generations);
            time2 = clock();
 printf("\nMOEA running time : %f \n",(float)(time2-
time1)/CLK_TCK*1000);
            OUTFILE(new_num_population,new_population);
 getch();
     return 0;
}
//-------------------------------------------------------
```

Input File (mf_MOEA.in)

```
14 40 1 2 /* Nodes Links Number of Destination Number of
flows */
13        /* Destination node */
512       /* Demand of the flow 1  */
512       /* Demand of the flow 2  */
0 1 1536 9
0 2 1536 9
0 3 1536 7
1 0 1536 9
2 0 1536 9
3 0 1536 7
1 3 1536 13
1 6 1536 20
3 1 1536 13
6 1 1536 20
2 4 1536 7
2 7 1536 26
4 2 1536 7
7 2 1536 26
3 10 1536 15
```

```
10  3  1536  15
4  5  1536  7
4  10  1536  11
5  4  1536  7
10  4  1536  11
5  6  1536  7
6  5  1536  7
6  9  1536  7
9  6  1536  7
7  8  1536  5
8  7  1536  5
7  13  1536  8
13  7  1536  8
8  9  1536  5
9  8  1536  5
8  12  1536  7
12  8  1536  7
9  11  1536  8
11  9  1536  8
10  12  1536  14
12  10  1536  14
10  11  1536  9
11  10  1536  9
12  13  1536  4
13  12  1536  4
```

Example 1

Number of flows: 1 (**512 Kbps**)

Destination node: 12

SOURCE	0
TARGET	12

		FUNCTIONS		
SOL	CHROMOSOME	HOP	DELAY	BAND. COMP
1	0,3,10,12	3	36	1536

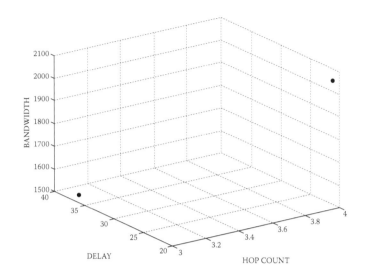

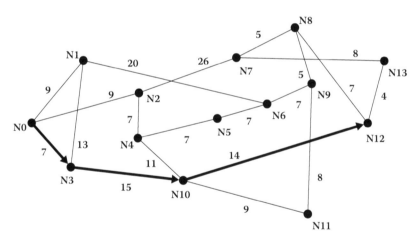

Example 2

Number of flows: 1 (512 Kbps)

Destination node: 13

SOURCE	0
TARGET	13

		FUNCTIONS		
SOL	CHROMOSOME	HOP	DELAY	BAND. COMP
1	0,2,7,13	3	43	1536
2	0,3,10,12,13	4	40	2048

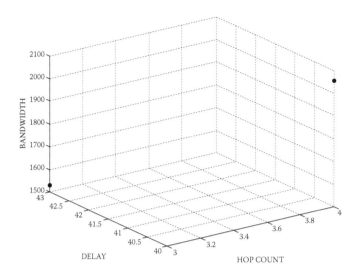

First solution:

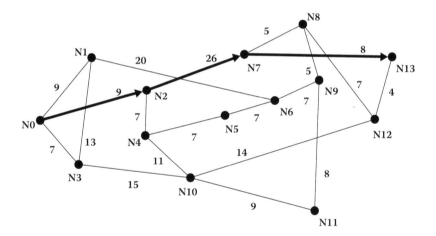

Second solution:

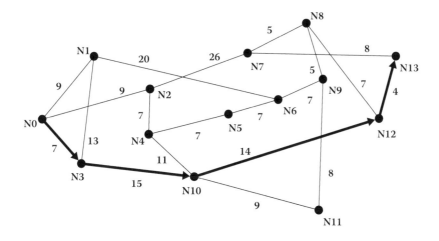

Example 3

Number of flows: 2 (512 Kbps every 1)

Destination node: **12**

SOURCE	0
TARGET	12

SOL	CHROMOSOME	FUNCTIONS		
		HOP	DELAY	BAND. COMP
	0,3,10,12			
1	0,3,10,12	6	72	3072

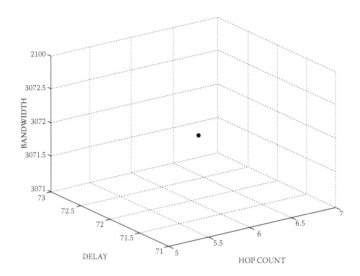

Solution:

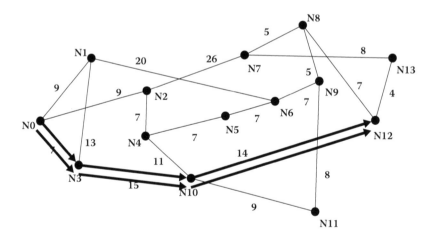

Example 4

Number of flows: 2 (512 Kbps every 1)

Destination node: 13

SOURCE	0
TARGET	13

SOL	CHROMOSOME	FUNCTIONS		
		HOP	DELAY	BAND. COMP
1	0,2,7,13			
	0,2,7,13	6	86	3072
2	0,3,10,12,13			
	0,3,10,12,13	8	80	4096
3	0,2,7,13			
	0,3,10,12,13	7	83	3584

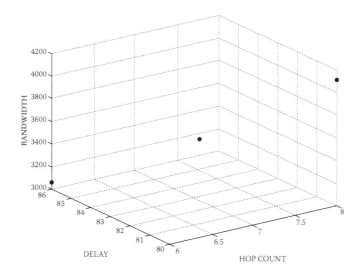

First solution:

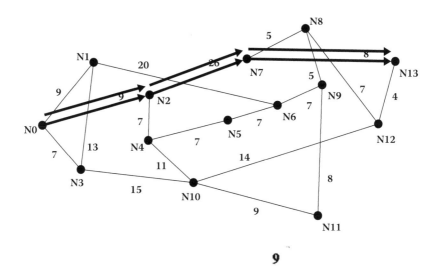

9

Second solution:

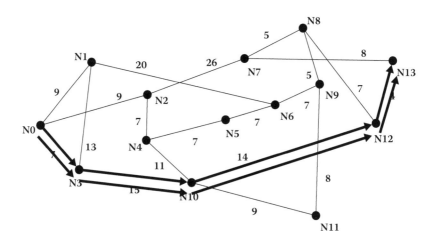

Third solution:

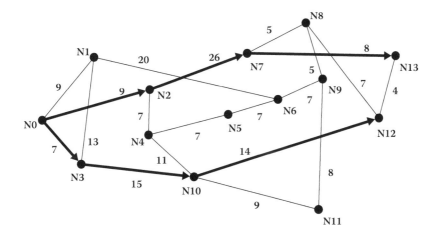

Index

Printed and bound by CPI Group (UK) Ltd, Croydon, CR0 4YY

17/10/2024

01775690-0018